The Accredited Symbian Developer Primer

The Accredited Symbian Developer Primer

Fundamentals of Symbian OS

By

Jo Stichbury and Mark Jacobs

Reviewed by

Jehad Al-Ansari, Rosanna Ashworth-Jones, Warren Day, Ioannis Douros, Graeme Duncan, Adrian Issot, Ashvin Lad, Rick Martin, Lorin McKay, Chris Notton, John Pagonis, Antony Pranata, William Roberts, Rahul Singh, Attila Vamos, Jonathan Yu

Symbian Press

Head of Symbian Press

Philip Northam

Managing Editor

Freddie Gjertsen

John Wiley & Sons, Ltd

Telephone (+44) 1243 779777

Email (for orders and customer service enquiries): cs-books@wiley.co.uk
Visit our Home Page on www.wiley.com

Other Wiley Editorial Offices

John Wiley & Sons Inc., 111 River Street, Hoboken, NJ 07030, USA

Jossey-Bass, 989 Market Street, San Francisco, CA 94103-1741, USA

Wiley-VCH Verlag GmbH, Boschstr. 12, D-69469 Weinheim, Germany

John Wiley & Sons Australia Ltd, 42 McDougall Street, Milton, Queensland 4064, Australia

John Wiley & Sons (Asia) Pte Ltd, 2 Clementi Loop #02-01, Jin Xing Distripark, Singapore 129809

John Wiley & Sons Canada Ltd, 6045 Freemont Blvd, Mississauga, ONT, L5R 4J3, Canada

Wiley also publishes its books in a variety of electronic formats. Some content that appears in print may not be available in electronic books.

British Library Cataloguing in Publication Data

A catalogue record for this book is available from the British Library

ISBN-13: 978-0-470-05827-5
ISBN-10: 0-470-05827-7

Typeset in 10/12pt Optima by Laserwords Private Limited, Chennai, India
Printed and bound in Great Britain by Bell & Bain, Glasgow
This book is printed on acid-free paper responsibly manufactured from sustainable forestry in which at least two trees are planted for each one used for paper production.

Contents

Foreword

Lee Epting, Vice President Developer Relations, Forum Nokia

The capabilities and features of mobile phones are expanding, and in consequence the software platforms upon which such phones are based are increasing in features and complexity. Further, consumers are adopting mobile software applications and services more broadly and thus the demand for quality is simultaneously increasing. For mobile phone software manufacturers to succeed, be they device manufacturers, independent software houses, or hobbyists, a key success factor today, and for the future, is high quality.

It is vital for software companies to recruit software engineers who truly understand the operating system for which they develop. Assessing such technical competence in an interview situation is not easy. The 'badge' Accredited Symbian Developer (ASD) is a form of proof of an individual's level of expertise, and thus recruiters can be confident that the person is a highly skilled Symbian C++ developer.

If an individual is not an Accredited Symbian Developer, the ASD exam provides real information about an individual's competencies. The Accredited Symbian Developer scheme facilitates the interview process.

If you are a developer wishing to extend your Symbian C++ expertise and become an Accredited Symbian Developer, there are many sources of valuable information from which you can learn. For example, developer websites such as ***www.forum.nokia.com*** and ***www.symbian.com/developer***, the discussion boards associated with these sites, and also hands-on training supplied both by Forum Nokia and by Symbian.

Books have long been the most widely-used way to acquire new programming skills; even today they hold their ground. Symbian Press has published a wide selection of books which cover various aspects of Symbian software development. For example, Jo Stichbury's previous Symbian Press book, *Symbian OS Explained*, is now widely used around the world both by developers and academics (as the basis for mobile software programming courses). I am sure that this, her second book, will prove equally invaluable to all who read it. Its premise is simple: if you understand this book you are sufficiently knowledgeable to become an Accredited Symbian Developer. The book delves into each of the topics contained within the ASD exam and explains them in sufficient detail to allow the reader to pass it.

Elisabet Melin, Vice President Marketing, UIQ Technology

I have worked with Symbian OS for eight years, first at Ericsson Mobile Communications, then at Symbian AB and eventually at UIQ Technology. In this time I have witnessed the Symbian community expand to the point where Symbian OS is now the leading operating system for advanced mobile phones, with the largest market share. This growth has created demand for competent software developers with a good rounded knowledge of Symbian C++ and its large number of APIs.

This book guides developers in their search for comprehensive knowledge and understanding of important concepts of Symbian OS, and prepares them for the Accredited Symbian Developer (ASD) exam. By helping developers to discover areas of Symbian OS they know less about than they'd like to, it provides an opportunity to acquire a solid base of knowledge from which to explore more specialist, niche areas.

UIQ Technology, as well as phone manufacturers and independent software houses, regularly recruits software engineers. We are, of course, looking for Symbian OS experts, developers who can create high quality software that will be running in phones, often 24/7, where both memory use and power consumption must be optimized.

A difficulty all recruiters face is that anyone can claim to be a Symbian OS expert. This primer and the ASD program are making it easier for developers to become Symbian OS experts and for companies in the Symbian OS community to recruit competent engineers. At UIQ Technology, we endorse the ASD program and support activities that put developers in our community in an optimal position to increase their competence and reach accredited status. Our developer program aims to cover the areas of the ASD exams when providing content and support. As there's much demand for experienced Symbian OS developers, getting accredited will give developers an edge with potential employers. Profiling based on accreditation is an easy way for an organization to make a good hiring decision.

The authors, Jo and Mark, have both worked at Symbian. Jo then pursued her career first at Sony Ericsson and then at Nokia – and managed to write *Symbian OS Explained* at the same time. Mark was involved in the development of the ASD exam and started a company in Canada to support ASD in North America. This combination of a vast development experience, deep knowledge of Symbian OS and teaching makes them the perfect authors for this primer, which I expect to become a classic.

When you can open this book at any page and already comprehend the topic, you will have the confidence that you're ready to tackle most Symbian OS projects.

About this Book

The Accredited Symbian Developer (ASD) Examination is fundamentally based on the content of existing Symbian Press books, with the C++ curriculum deriving from general C++ literature. Arguably, if you own the core Symbian Press titles and a solid C++ reference book, you will have all the information you require to get through the exam. So why do you need an ASD Primer?

The problem is that there is a lot to know! Although the exam objectives are published, it can still be unclear what is really relevant and what is not.

To address this, the Primer is a pragmatic restructuring of a number of existing Symbian Press titles, primarily Jo's *Symbian OS Explained: Effective C++ Programming for Smartphones*. It also borrows from Richard Harrison's *Symbian OS C++ for Mobile Phones, Volume 2* and Steve Babin's *Developing Software for Symbian OS*. Where necessary, material from those books has been updated to meet the needs of an audience working with Symbian OS v9.1 and beyond. To do this, we took guidance on Platform Security and EKA2 from Craig Heath's *Symbian OS Platform Security* and Jane Sales' *Symbian OS Internals,* respectively.

We have organized the information in such a way that the reader can dip in and out of sections to refresh their knowledge and understanding of the fundamental concepts of Symbian software development. In this sense, the Primer can also serve as a desk reference for both Symbian OS programming and, more modestly, C++, as well as a revision guide for ASD exam candidates.

The C++ sections have focused on information in Stroustrop's essential *The C++ Programming Language* as the primary reference, but we also found ourselves returning to Scott Meyer's invaluable *Effective C++* and

Stephen Dewhurst's excellent *C++ Common Knowledge*. Please see the References section for more information.

Our intent was not to rehash material but to provide every ASD candidate with a solid foundation in the subject matter, so that they can enter the exam with confidence.

That said, simply reading the book, or even having it present when you sit the exam, will not guarantee you a pass, because you will still need to understand the subject matter. The best classroom for this is the workplace or a study project, where the lessons learnt in this book can be put into practice. If you understand the contents of this Primer and have about two years' exposure to Symbian OS, you should have the requisite skills to pass with flying colors and become an ASD.

Example ASD questions and the sample code for this book are available for download from the Symbian Press website, ***www.symbian.com/developer/books***, and from Meme Education, ***www.meme-education.com***.

We hope you enjoy the journey towards Accreditation!

Jo Stichbury
Mark Jacobs
Summer 2006

About the Authors

Jo Stichbury

Jo Stichbury was educated at Magdalene College, Cambridge, where she held the Stothert Bye-Fellowship. She has an MA in Natural Sciences and a PhD in the chemistry of Organometallic Molybdenum complexes. She has worked within the Symbian ecosystem since 1997 in the Base, Connectivity and Security teams of Symbian. She has also worked for Advansys, Sony Ericsson and Nokia. Jo became an Accredited Symbian Developer in 2005 and a Forum Nokia Champion in 2006.

Jo is also the sole author of *Symbian OS Explained: Effective C++ Programming for Smartphones* which was published by Symbian Press in 2004. She fled the UK for Canada shortly before it was released, but was tracked down in Vancouver and pressed back into service to co-author this new book about the Accredited Symbian Developer exams.

Mark Jacobs

Mark wrote his first computer program in early 1980. He wrote his first C++ program in 1987, in an effort to prove that Modula-2 was the better language, and has enjoyed an ongoing bipolar relationship with the language ever since.

He joined Symbian in January 2000, stage diving into the last few months of development of Symbian OS v6.0, which was not totally dissimilar to a Motorhead back-stage party hosted by NASA. For the majority of his time at Symbian, Mark was a System Architect. He left London for Vancouver in 2004 and founded Meme Education in late 2005.

He has a BSc in Computer Science from the University of Hertfordshire and is an Accredited Symbian Developer. He lives with Jo, two Siamese cats and a La Pavoni espresso machine.

Acknowledgments

First and foremost, a big thank you to Coach Freddie (Sven) Gjertsen of Symbian Press. Freddie has been a complete star: listened to our rants, bribed us, and just got stuff done.

No Symbian Press book would be complete without a mention of Phil Northam, who came to Vancouver on what can only be described as a 5500 mile, first-class, extended pub crawl of the Piccadilly Line, to buy us lunch and to convince Jo that she desperately needed to write another book. Thanks also to Drew Kennerly and all at John Wiley for making things worryingly easy, and to Satu McNabb of Symbian Press.

A note of thanks goes to our fellow Symbian Press authors: Richard Harrison, Steve Babin, Craig Heath, Jane Sales (and Boris and Mishka), Michael Jipping and all those who worked on the fine Symbian Press titles that we drew upon while writing this one.

We'd also like to thank the reviewers who contributed enormously to the technical accuracy of this book.

Jo would like to thank her colleagues for their support and understanding while she took time out to write this book, in particular Van Ly and Jon Bruce, Bill Bonney, Amonn Phillip and Kevin Chan. Thank you for your patience; normal service should be resumed once I figure out where I left it.

A special thanks goes to Ian Weston of Majinate, whose integrity, advice and just plain "doing the right thing" approach has been invaluable to us. A nod of respect also goes to the good folk of Symsource. And, of course all those who wrote the exam. Thanks, chaps – they still hurt even when you know the answers.

This book was written using Pages 2. Thanks to Apple Computer, Inc. for giving us an alternative.

Peace and love to James, Yvonne, Viv, Val, Nicky, Clive, and Scott.

Symbian Press Acknowledgements

We would like to thank Mark and Jo for the long hours that they dedicated to the creation of this book and hope that they are not too emasculated by their endeavors.

Thanks are also due to all our captious reviewers, named and otherwise, for the insights that they produced, to deadline.

Introduction

by Ian Weston, Majinate

Software development on the Symbian platform is a delicate discipline, requiring an appreciation of a wide range of issues that do not necessarily apply to other development environments. Symbian OS, since its creation, has been specifically developed for mobile phones, and as a consequence it uses different development concepts from that of standard desktop or server development. These concepts are designed to:

- provide efficient power management
- ensure fast boot times
- allow phones to run indefinitely without rebooting
- conserve and recycle memory
- allow software installation without restarting the operating system
- make use of the permanent connectedness of networked devices
- make phones robust when networks and connections fail.

What Is the Accredited Symbian Developer Scheme?

The balance of opportunity and constraints is clearly different for mobile software development, and a software developer who appreciates and understands these Symbian OS concepts is a highly skilled individual. The Accredited Symbian Developer (ASD) qualification allows professional Symbian developers to validate their understanding and knowledge of Symbian OS software development with an industry-recognized certificate of professional achievement. The goal of the ASD scheme is to ensure that the Symbian ecosystem is provided with better-educated developers, and to allow those developers to differentiate themselves within the ecosystem.

The ASD scheme publishes the "ASD Curriculum", the set of knowledge required to understand and develop software for Symbian OS. The scheme, and especially the curriculum, provides a framework for continuing professional development, keeping pace with the evolution of Symbian OS. The curriculum is updated as new features are introduced into the operating system and old features are deprecated. This is done in a measured manner with full visibility of the future use of Symbian OS in mobile phones and with knowledge of the long-term evolution of the system.

The curriculum ensures that exam candidates are tested objectively, on material that is readily accessible. The exam is able to differentiate between levels of proficiency to the extent that candidates above a threshold of competence, independently defined and reviewed by Symbian, are assigned the status "Accredited Symbian Developer".

The ASD infrastructure provides an excellent tool for unbiased, non-threatening testing of competency and understanding of the curriculum. The option of a detailed proficiency analysis allows training managers to target training investment appropriately and assess the return accurately through a sequence of exams giving measurable improvements in performance. In turn, this means that employers can better allocate engineering staff to appropriate tasks and roles. The full examination can also be used to obtain interview preparation sheets by generating a report of the relative strengths and weaknesses of the candidate across the curriculum topics. This is highly valuable in interview screening, giving more effective use of face-to-face interview time and avoiding the need for longer technical interviews.

The Accredited Symbian Developer Exam in Detail

The examination infrastructure is web-based and supplied by Majinate Limited, enabling the exam to be sat anywhere in the world with an internet connection by visiting ***www.majinate.com/takeexam***.

The exam questions test the candidate's knowledge of each of the ASD curriculum areas. These questions range in difficulty from straightforward (requiring little real-world exposure and limited book study) through to devilishly hard (requiring years of experience and a thorough understanding of the principles of the operating system). During any examination session, an adaptive mechanism, which takes account of how well the previous questions on a subject were answered, is used to determine the difficulty of the next question. This ensures that each candidate is presented with questions that test to the limits of his or her understanding.

The examination does not have a predetermined number of questions nor duration; however, the majority of candidates can expect to complete the test in 70 minutes and spend approximately 90 seconds on each question.

Each question in the exam is presented as a statement with five answers. Up to three of these answers may be correct. The candidate is required to mark as many of the correct answers as possible and avoid marking any wrong answers (which may be there to distract the unwary). Only by selecting all the correct answers in the fastest time will the maximum number of marks for a question be achieved. Negative marks are awarded for slowness, for failing to select a right answer, or for selecting a wrong one. Skipped questions (those where no answers are chosen by the candidate) are marked conservatively, and a single choice of answer which is wrong may lead to a better final mark than a skipped question. In deciding how to answer a question, therefore, candidates should try to choose at least one answer which is likely to be right.

Exam Results and Controls

At the end of the exam, the results are analyzed offline and compared with the Symbian-defined pass mark. Within 48 hours, the candidate receives an email with the exam result. Candidates who pass are informed of their certificate number and when the certificate will be delivered to them. Candidates who do not demonstrate the required standard are advised of the areas where they were weakest as a recommendation for further study. There is no cooling-off period, and a candidate is free to re-sit the exam at any time. There is no consideration of the earlier exam in the presentation, analysis or results of later attempts.

In order to determine the level a candidate must attain to be recognized as an ASD, the initial pass mark was determined by examining a large contingent of Symbian and Symbian Partner engineers. The pass mark is reviewed annually to ensure that it maintains the high standards of the ASD scheme.

There are built-in security features to detect anomalous behavior that might indicate cheating, and the system reacts accordingly. In order to maximize the accessibility of the exam, it can be run in two modes: for candidates who are unable or unwilling to attend a testing centre supervised by one of Majinate's partner companies worldwide, it is possible to sit the exam in self-supervised mode. Self-supervised candidates sit exactly the same exam as supervised candidates (other than the exam environment) and receive a certificate indicating that the exam was not taken under supervision. Majinate is unable to confirm that a person holding such a certificate did indeed sit the exam unaided. For supervised candidates this is guaranteed by the supervisor, whose details are supplied with the examination certificate.

Potential employers and interviewers are able to run a short form of the exam online and on demand, with results available immediately. This, although not sufficient to repeat the examination exercise, confirms whether the candidate falls into the correct percentile as claimed by the certificate.

Exam Essentials Summary

1 C++ Language Fundamentals

1.1 Types

- Understand that C++ has a number of different types in various categories for example integral, arithmetic
- Recognize `typedef` as defining a synonym for an existing type but not a new type in itself
- Recognize enumeration types as user-defined value sets
- Specify the advantages of `const` over `#define`
- Understand the use and properties of the C++ reference type
- Specify the difference between pointers and references
- Understand the semantics of pointer arithmetic
- Recognize pointer operations and the purpose of the `NULL` pointer value
- Differentiate `const` pointers and pointers to `const`

1.2 Statements

- Know the use and properties of the declaration and definition statements
- Recognize the use of the `extern` keyword

- Cite and understand initialization of variables and their scope
- Understand the purpose, syntax and behavior of C++ loop statements (`while`, `for` and `do`)
- Specify the behavior and effect of the `continue` and `break` keywords in a loop
- Specify the syntax and behavior of C++ conditional statements (`if` and `switch`)

1.3 Expressions and Operators

- Specify the syntax and meaning of unary and binary operator expressions
- Understand the difference between precedence and associativity
- Recognize common operator categories including logical, prefix and postfix
- Demonstrate awareness of general operator precedence and associativity rules

1.4 Functions

- Understand the syntax of a function prototype
- Cite the purpose of the `inline` keyword
- Understand the rules for passing default arguments and an unspecified number of arguments to a function
- Recognize value, reference, array and pointer parameter passing
- Understand the scope of function blocks and return by reference and value
- Specify the syntax for pointers to function assignments
- Recognize pointer to functions as callback parameter arguments

1.5 Dynamic Memory Allocation

- Understand C++ free store allocation scope using the `new` and `delete` operators
- Recognize the syntax and purpose of placement `new`

1.6 Tool Chain Basics

- Understand the function of the tools, the C++ tool chain (for example compiler and linker)
- Recognize the lexical and syntax-parsing stages of compilation
- Be able to describe the purpose of the C++ preprocessor, specifying common directives
- Understand the role `inline` functions play in C++
- Know how to use the `extern` keyword

2 Classes and Objects

2.1 Scope and C++ Object-Oriented Programming (OOP) Support

- Understand the scope and lifetime properties of blocks and namespaces
- Understand C++ support for data abstraction
- Specify the attributes of a C++ object with respect to object-oriented programming
- Know the syntax of a class declaration
- Cite the differences between a class and an object
- Differentiate between basic data structures (`struct` keyword) and classes

2.2 Constructors and Destructors

- Understand the order of construction and destruction for the class and its member variables
- Recognize implicit constructor invocation (overloading and pattern matching) and the role the `explicit` keyword plays in constructor declaration
- Specify what the compiler automatically generates for user-defined classes
- Understand the purpose and use of copy constructors (including parameter passing)
- Understand the difference between assignment and initialization

- Specify the required function members needed in a class to support safe ownership of pointer data

2.3 Class Members

- Describe `private`, `protected` and `public` access control for class members
- Declare and specify the syntax for pointers to class members
- Identify nested classes and their scope and lifespan
- Cite the scope access rules and lifetime of a nested class
- Understand the scope rules and syntax of `friend` functions
- Understand the semantics of member functions and data addressing
- Understand the purpose of the scope-resolution operator `::`
- Understand the role of the `this` pointer
- Specify the properties of `static` class members

3 Class Design and Inheritance

3.1 Class Relationships

- Understand the key benefits and purpose of inheritance
- Specify the differences between composition, aggregation and inheritance
- Cite the object-oriented relationship for inheritance

3.2 Inheritance

- Be able to define public inheritance
- Understand the scope resolution operator syntax for accessing the base-derived class hierarchy
- Given a base-derived hierarchy, specify the access rules (including those of friend classes)
- Describe the scope access rules and purpose of public, protected and private inheritance
- Specify the implicit invocation order of constructors and destructors in a base-derived hierarchy

3.3 Dynamic Polymorphism – Virtual Methods

- Specify the mechanisms of OO reuse available in C++
- Be to able to state C++ support for polymorphism
- Understand the purpose of and difference between overriding and overloading
- Understand the use of overriding to modify behavior in base-derived class inheritance
- Understand the rules and pattern-matching criteria for correct overloaded-function invocation
- Identify the typical uses and behavior for operator overloading
- Describe the purpose of the virtual table, citing constraints and overheads
- Specify the use of virtual functions and their implementation tradeoffs
- Understand how an abstract base class is implemented in C++
- State the differences between interface and implementation inheritance
- Understand and recognize the problems associated with multiple inheritance
- Cite the implementation requirements to support the `static_cast` operator in user-defined classes

3.4 Static Polymorphism and Templates

- Specify the syntax for a simple function template specialization
- Be able to cite the advantages of function templates (for example over macros)
- Understand the inheritance rules and syntax supported by class templates
- Understand the syntax and semantics of a template type/class declaration
- Recognize the prototype declaration and pattern-matching properties of a template declaration and its use
- Understand the purpose and implementation differences of the Symbian OS thin template and mainstream C++ templates

4 Symbian OS Types and Declarations

4.1 The Fundamental Symbian OS Types

- Know how the fundamental Symbian OS types relate to native built-in C++ types
- Understand that the fundamental types should always be used in preference to the native built-in C++ types (`bool`, `int`, `float`, etc.) because they are compiler-independent

4.2 T Classes

- Know the purpose of a T class, what types of member data it may and may not own, and that it must never have a destructor
- Know what types of function a T class may have
- Understand that a T class may be created on the heap or stack
- Understand that a T class may be used as an alternative to the traditional C/C++ `struct`
- Know that the T prefix is also used to define an `enum`.

4.3 C Classes

- Recognize that a C class always derives from CBase
- Know the purpose of a C class, and what types of data it may own
- Understand that a C class must always be instantiated on the heap
- Know that a C class uses two-phase construction and has its member data zero-filled when it is allocated on the heap
- Understand the destruction of C classes via the virtual destructor defined in `CBase`

4.4 R Classes

- Know the purpose of an R class, to own a resource
- Understand that an R class can be instantiated on the heap or the stack
- Understand the separate construction and initialization of R classes
- Understand the separate cleanup and destruction of R classes, and the consequences of forgetting to call the `Close()` or `Reset()` method before destruction

4.5 M Classes

- Know the purpose of an M class, to define an interface
- Understand the use of M classes for multiple inheritance, and the order in which to derive an implementation class from C and M classes
- Know that an M class should never contain member data and does not have constructors
- Know what types of function an M class may include and the circumstances where it is appropriate to define their implementation
- Understand that an M class cannot be instantiated

4.6 Static Classes

- Know that static classes do not have a prefix letter
- Understand that static classes cannot be instantiated because they contain only static functions

4.7 Factors to Consider when Creating a Symbian OS Class

- Know the important factors to consider when creating a new class, and how this determines the choice of Symbian OS class type

4.8 Why Is the Symbian OS Naming Convention Important?

- Understand that the use of a class prefix makes it clear to anyone wishing to use a class how it should be instantiated, used and destroyed safely.
- Recognize that the naming convention forces a class designer to think about the factors described in Section 4.7 and, having decided on the fundamental behavior, can concentrate on the role of the class, knowing that leave-safe construction, destruction and ownership are already handled.

5 Leaves and the Cleanup Stack

5.1 Leaves: Lightweight Exceptions for Symbian OS

- Know that, before v9, Symbian OS does not support standard C++ exceptions (`try/catch/throw`) but uses a lightweight alternative: `TRAP` and leave

- Know that leaves are a fundamental part of Symbian error handling and are used throughout the system
- Understand the similarity between leaves and the `setjmp/longjmp` declarations in C
- Recognize the typical system functions which may cause a leave, including the `User::LeaveXXX()` functions and `new(ELeave)`
- Be able to list typical circumstances which cause a leave (for example, insufficient memory for a heap allocation)
- Understand that `new(ELeave)` guarantees that the pointer return value will always be valid if a leave has not occurred

5.2 How to Work with Leaves

- Know that leaves are indicated by use of a trailing L suffix on functions containing code which may leave (for example, `InitializeL()`)
- Be able to spot functions which are not leave-safe and those which are
- Understand that leaves are used for error handling; code should very rarely both return an error and be able to leave
- Understand the reason why a leave should not occur in a constructor or destructor

5.3 Comparing Leaves and Panics

- Understand the difference between a leave and a panic
- Recognize that panics come about through assertion failures, which should be used to flag programming errors during development
- Recognize that a leave should not be used to direct normal code logic

5.4 What Is a `TRAP`?

- Recognize the characteristics of a `TRAP` handler
- Understand that, for efficiency, use of `TRAP`s should be kept to a minimum

5.5 The Cleanup Stack

- Know how to use the cleanup stack to make code leave-safe, so memory is not leaked in the event of a leave

- Understand that `CleanupStack::PushL()` will not leak memory even if it leaves
- Know the order in which to remove items from the cleanup stack, and how to use `CleanupStack::PopAndDestroy()` and `CleanupStack::Pop()`
- Recognize correct and incorrect use of the cleanup stack
- Understand the consequences of putting a C class on the cleanup stack if it does not derive from `CBase`
- Know how to use `CleanupStack::PushL()` and `CleanupXXXPushL()` for objects of C, R, M and T classes and `CleanupArrayDeletePushL()` for C++ arrays
- Understand the meaning of the Symbian OS function suffixes C and D

5.6 Detecting Memory Leaks

- Recognize the use of the `__UHEAP_MARK` and `__UHEAP_MARKEND` macros to detect memory leaks

6 Two-Phase Construction and Object Destruction

6.1 Two-Phase Construction

- Know why code should not leave inside a constructor
- Recognize that two-phase construction is used to avoid the accidental creation of objects with undefined state
- Understand that constructors and second-phase `ConstructL()` methods are given private or protected access specifiers in classes which use two-phase construction, to prevent their inadvertent use
- Understand how to implement two-phase construction, and how to construct an object which derives from a base class which also uses a two-phase method of initialization
- Know the Symbian OS types (C classes) which typically use two-phase construction

6.2 Object Destruction

- Know that it is neither efficient nor necessary to set a pointer to `NULL` after deleting it in destructor code

- Understand that a destructor must check before dereferencing a pointer in case it is `NULL`, but need not check if simply calling `delete` on that pointer

7 Descriptors

7.1 Features of Symbian OS Descriptors

- Understand that Symbian OS descriptors may contain text or binary data
- Know that descriptors may be narrow (8-bit), wide (16-bit) or neutral (which is 16-bit since Symbian OS is built for Unicode)
- Understand that descriptors do not dynamically extend the data area they reference, so will panic if too small to store data resulting from a method call

7.2 The Symbian OS Descriptor Classes

- Know the characteristics of the `TDesC`, `TDes`, `TBufC`, `TBuf`, `TPtrC`, `TPtr`, `RBuf` and `HBufC` descriptor classes
- Understand that the descriptor base classes `TDesC` and `TDes` implement all generic descriptor manipulation code, while the derived descriptor classes merely add construction and assignment code specific to their type
- Identify the correct and incorrect use of modifier methods in the `TDesC` and `TDes` classes
- Recognize that there is no `HBuf` class, but that `RBuf` can be used instead as a modifiable dynamically allocated descriptor

7.3 The Inheritance Hierarchy of the Descriptor Classes

- Know the inheritance hierarchy of the descriptor classes
- Understand the memory efficiency of the descriptor class inheritance model and its implications

7.4 Using the Descriptor APIs

- Understand that the descriptor base classes `TDesC` and `TDes` cannot be instantiated

- Understand the difference between `Size()`, `Length()` and `MaxLength()` descriptor methods
- Understand the difference between `Copy()` and `Set()` descriptor methods and how to use assignment correctly

7.5 Descriptors as Function Parameters

- Understand that the correct way to specify a descriptor as a function parameter is to use a reference, for both constant data and data that may be modified by the function in question.

7.6 Correct Use of the Dynamic Descriptor Classes

- Identify the correct techniques and methods to instantiate an `HBufC` heap buffer object
- Recognize and demonstrate knowledge of how to use the new descriptor class `RBuf`

7.7 Common Inefficiencies in Descriptor Usage

- Know that `TFileName` objects should not be used indiscriminately, because of the stack space each consumes
- Understand when to dereference an `HBufC` object directly, and when to call `Des()` to obtain a modifiable descriptor (`TDes&`)

7.8 Literal Descriptors

- Know how to manipulate literal descriptors and know that those specified using `_L` are deprecated
- Specify the difference between literal descriptors using `_L` and those using `_LIT` and the disadvantages of using the former

7.9 Descriptor Conversion

- Know how to convert 8-bit descriptors into 16-bit descriptors and vice versa using the descriptor `Copy()` method or the `CnvUtfConverter` class
- Recognize how to read data from file into an 8-bit descriptor and then 'translate' the data to 16-bit without padding, and vice versa

- Know how to use the `TLex` class to convert a descriptor to a number, and `TDes::Num()` to convert a number to a descriptor

8 Dynamic Arrays

8.1 Dynamic Arrays in Symbian OS

- Demonstrate an understanding of the basics of Symbian OS dynamic arrays (`CArrayX` and `RArray` families)
- Understand the different types of Symbian OS dynamic arrays with respect to memory arrangement (flat or segmented), object storage (within array or elsewhere), object length (fixed or variable) and object ownership.
- Recognize the appropriate circumstances for using a segmented-buffer array class rather than a flat array class

8.2 `RArray`, `RPointerArray` or `CArrayX`?

- Know the reasons for preferring `RArrayX` to `CArrayX`, and the exceptional cases where `CArrayX` classes are a better choice

8.3 Array Granularities

- Understand the meaning of array granularity and capacity
- Know how to choose the granularity of an array as appropriate to its intended use

8.4 Array Sorting and Searching

- Demonstrate an understanding of how to sort and seek in dynamic arrays
- Recognize that `RArray`, `RPointerArray` and the `CArrayX` family can all be sorted, although the `CArrayX` classes are not as efficient

8.5 `TFixedArray`

- Recognize that, when a dynamic array is not required, the `TFixedArray` class should be preferred over a C++ array, since it gives the benefit of bounds checking (debug-only or debug and release)

9 Active Objects

9.1 Event-Driven Multitasking on Symbian OS

- Demonstrate an understanding of the difference between synchronous and asynchronous requests and be able to differentiate between typical examples of each
- Recognize the typical use of active objects to allow asynchronous tasks to be requested without blocking a thread
- Understand the difference between multitasking using multiple threads and multiple active objects, and why the latter is preferred in Symbian OS code

9.2 Class `CActive`

- Understand the significance of an active object's priority level
- Recognize that the active object event handler method (`RunL()`) is non-pre-emptive
- Know the inheritance characteristics of active objects, and the functions they are required to implement and override
- Know how to correctly construct, use and destroy an active object

9.3 The Active Scheduler

- Understand the role and characteristics of the active scheduler
- Know that `CActiveScheduler::Start()` should only be called after at least one active object has an outstanding request
- Recognize that a typical reason for a thread to fail to handle events may be that the active scheduler has not been started or has been stopped prematurely
- Understand that `CActiveScheduler` may be sub-classed, and the reasons for creating a derived active scheduler class

9.4 Canceling an Outstanding Request

- Understand the different paths in code that the active object uses when an asynchronous request completes normally, and as the result of a call to `Cancel()`

9.5 Background Tasks

- Understand how to use an active object to carry out a long-running (or background) task
- Demonstrate an understanding of how self-completion is implemented

9.6 Common Problems

- Know some of the possible causes of stray signal panics, unresponsive event handling and blocked threads

10 System Structure

10.1 DLLs in Symbian OS

- Know and understand the characteristics of polymorphic interface and shared library (static) DLLs
- Know that UID2 values are used to distinguish between static and polymorphic DLLs, and between plug-in types
- For a shared library, understand which functions must be exported if other binary components are to be able to access them
- Know that Symbian OS does not allow library lookup by name but only by ordinal

10.2 Writable Static Data

- Recognize that writable static data is not allowed in DLLs on EKA1 and discouraged on EKA2
- Know the basic porting strategies for removing writable static data from DLLs

10.3 Executables in ROM and RAM

- Recognize the correctness of basic statements about Symbian OS execution of DLLs and EXEs in ROM and RAM

10.4 Threads and Processes

- Recognize the correctness of basic statements about threads and processes on Symbian OS

- Recognize the role and the characteristics of the synchronization primitives `RMutex`, `RCriticalSection` and `RSemaphore`

10.5 Inter-Process Communication (IPC)

- Recognize the preferred mechanisms for IPC on Symbian OS (client–server, publish and subscribe and message queues), and demonstrate awareness of which mechanism is most appropriate for given scenarios
- Understand the use of publish and subscribe to retrieve and subscribe to changes in system-wide properties, including the role of platform security in protecting properties against malicious manipulation

10.6 Recognizers

- Recognize correct statements about the role of recognizers in Symbian OS

10.7 Panics and Assertions

- Know the type of parameters to pass to `User::Panic()` and understand how to make them meaningful
- Understand the use of `__ASSERT_DEBUG` statements to detect programming errors in debug code by breaking the flow of code execution using a panic
- Recognize that `__ASSERT_ALWAYS` should be used more sparingly because it will test statements in released code too and cause code to panic if the assertion fails

11 Client–Server Framework

11.1 The Client–Server Pattern

- Know the structure and benefits of the client–server framework
- Understand the different roles of system and transient servers, and match the appropriate server type to examples of server applications

11.2 Fundamentals of the Symbian OS Client–Server Framework

- Know the fundamentals of the Symbian OS client–server implementation

11.3 Symbian OS Client–Server Classes

- Know the classes used by the Symbian OS client–server framework, and basic information about the role of each
- Recognize the objects that a server must instantiate when it starts up
- Understand the mechanism used to prevent the spoofing of servers in Symbian OS

11.4 Client–Server Data Transfer

- Know the basics of how clients and servers transfer data for synchronous and asynchronous requests
- Recognize the correct code to transfer data from a client derived from `RSessionBase` to a Symbian OS server
- Know how to submit both synchronous and asynchronous client–server requests
- Know how to convert basic and custom data types into the appropriate payload which can be passed to the server, as both read-only and read/write request arguments

11.5 Impact of the Client–Server Framework

- Understand the potential impact on run-time speed from using a client–server session and differentiate between circumstances where it is useful or necessary and where it is inefficient
- Recognize scenarios where an implementation which uses client subsessions with the server would be recommended
- Understand the impact of the context switch required when making a client–server request, and the best way to manage communication between a client and a server to maximize run-time efficiency

12 File Server and Streams

12.1 The Symbian OS File System

- Understand the role of the file server in the system
- Know the basic functionality offered by class `RFs`
- Recognize code which correctly opens a fileserver session (`RFs`) and a file subsession (`RFile`) and reads from and writes to the file

- Know the characteristics of the four `RFile` API methods which open a file
- Understand how `TParse` can be used to manipulate and query file names

12.2 Streams and Stores

- Know the reasons why use of the stream APIs may be preferred over use of `RFile`
- Understand how to use the stream and store classes to manage large documents most efficiently
- Be able to recognize the Symbian OS store and stream classes and know the basic characteristics of each (for example base class, memory storage, persistence, modification, etc.)
- Understand how to use `ExternalizeL()` and operator << with `RWriteStream` to write an object to a stream, and `InternalizeL()` and operator >> with `RReadStream` to read it back
- Recognize that operators >> and << can leave

13 Sockets

13.1 Introducing Sockets

- Recognize correct high-level statements which define and describe a network socket
- Recognize correct statements about transport independence
- Know the difference between connected and connectionless sockets
- Differentiate between streamed and datagram communication and their relationship with connected/connectionless sockets

13.2 The Symbian OS Sockets Architecture

- Demonstrate a basic understanding of the support for sockets on Symbian OS
- Recognize the characteristics of the `RSocketServ`, `RSocket` and `RHostResolver` classes
- Understand the role and purpose of PRT protocol modules

13.3 Using Symbian OS Sockets

- Recognize correct patterns for opening and configuring connected and connectionless sockets
- Know which `RSocket` API methods should be used for connected and unconnected sockets to send and receive data
- Know the characteristics of the synchronous and asynchronous methods for closing an `RSocket` subsession

14 Tool Chain

14.1 Build Tools

- Understand the basic use of `bldmake`, `bld.inf` and `abld.bat`
- Understand the purpose and typical syntax of project definition (MMP) files
- Understand the role of Symbian OS resource and text localization files

14.2 Hardware Builds

- Understand that the ARM C++ EABI is an industry standard optimized for embedded application development
- Recognize basic information about the RVCT and GCCE compilers, which can be used for target hardware builds
- Understand that ARMV5 supports both 32-bit ARM and 16-bit THUMB instructions, and appreciate the difference with respect to speed and size

14.3 Installing an Application to Phone Hardware

- Recognize the package file format used for creation of SIS installation files

14.4 The Symbian OS Emulator

- Understand the purpose of the Symbian OS emulator for Windows
- Recognize differences between running code on the emulator and on target hardware

15 Platform Security

15.1 The Trust Model

- Understand what is meant by the axiom "a process is a unit of trust" and how Symbian OS enforces this
- Understand the purpose of the Trusted Computing Base and why it is important
- Recognize that a number of Symbian OS APIs do not require security checks before they can be used
- Know that self-signed software that does not use sensitive system services is "untrusted" and can be installed and run on the phone, although it is effectively "sandboxed"

15.2 Capability Model

- Understand the relationship between capabilities and the Trusted Computing Base (TCB)
- Understand the concept of user capabilities and their relationship to the Trusted Computing Environment (TCE)
- Understand the relationship between the TCB/TCE, capability assignment, software install as the "gatekeeper" and the role of application signing
- Recognize the different groups of capabilities, demonstrating a broad understanding of the privileges granted
- Recognize how to specify platform security capabilities within an MMP file
- Demonstrate an understanding of the capability rules

15.3 Data Caging

- Understand how data caging works to protect all types of files via the three special directories (`\sys`, `\resource` and `\private`); in particular, that data caging is used to partition all executables in the file system so, once trusted, they are protected from modification
- Understand the implications of data caging for naming executable code
- Recognize that data caging can be used to provide a secure area for an application's data

- Recognize the capabilities needed to read from and write to specific directories and subdirectories
- Know that DLLs do not have a private data-caged area and use that of the process in which they are loaded, and that this directory can be acquired by the DLL using the file system methods `RFs::PrivatePath()`

15.4 Secure Identifier, Vendor Identifier and Unique Identifier

- Explain what a Secure Identifier (SID) is, where it is defined and what it is used for
- Understand the similarities and differences between a Secure Identifier (SID), a Vendor Identifier (VID) and a binary's Unique Identifiers (UID)
- Know the rules by which an application is identified, according to the specification of SID, VID and UID
- Understand that SID and VID may be assigned, but are not relevant, to DLLs
- Recognize how to specify VID and SID within an MMP file
- Understand that UIDs are now split into 2 groups (protected and unprotected ranges) with different implications for test and commercial code

15.5 Application Design for a Secure Platform

- Demonstrate an understanding of the key considerations when writing a secure application, including the parties interested in application security, typical attacks, countermeasures and secure application design, and the costs of various countermeasures

15.6 Releasing a Secure Application on Symbian OS v9

- Understand the basic process of testing and releasing a signed V9 application

15.7 The Native Software Installer

- Recognize the key functions of the v9 Native Software Installer, including the compatibility break in SIS file format between v9 and previous versions of Symbian OS

16 Compatibility

16.1 Levels of Compatibility

- Demonstrate an understanding of source, binary, library, semantic and forward/backward compatibility

16.2 Preventing Compatibility Breaks – What Cannot Be Changed?

- Recognize which attributes of a class are necessary for a change in the size of the class data not to break compatibility
- Understand which class-level changes will break source compatibility
- Understand which class-level changes will break binary compatibility
- Understand which library-level changes will break binary compatibility
- Understand which function-level changes will break binary and source compatibility
- Differentiate between derivable and non-derivable C++ classes in terms of what cannot be changed without breaking binary compatibility

16.3 What Can Be Changed Without Breaking Compatibility?

- Understand which class-level changes will not break source compatibility
- Understand which class-level changes will not break binary compatibility
- Understand which library-level changes will not break binary compatibility
- Understand which function-level changes will not break binary and source compatibility
- Differentiate between derivable and non-derivable C++ classes in terms of what can be changed without breaking binary compatibility

16.4 Best Practice – Designing to Ensure Future Compatibility

- Recognize best practice for maintaining source and binary compatibility
- Recognize the coupling arising from the use of inline functions and differentiate between cases where it will make maintaining binary compatibility more difficult and where it will be less significant

1

C++ Language Fundamentals

Introduction

The creator of C++, Bjarne Stroustrup, states: "C++ is a general-purpose programming language ... that is a better C, supports data abstraction, supports object-oriented programming and supports generic programming." [Stroustrup 2000] It was because of these qualities, in particular object-oriented programming support, that in 1994 Symbian (then part of Psion) adopted C++ when rewriting their 16-bit operating system SIBO to create EPOC32, which has since evolved into Symbian OS.

Symbian OS C++ is built on top of C++ with some deliberate omissions, including exceptions, templates and the standard template library. It also extends or, more accurately, complements C++ by providing strong coding conventions, small-memory management techniques and highly optimized algorithms for mobile phones. To program in C++ on Symbian OS, developers require strong skills in mainstream C++.

This chapter covers the required language basics for the ASD exam: types, statements, functions and a little about the tool chain. It touches on some object-oriented (OO) syntax but only where needed, as this is dealt with in more detail in Chapters 2 and 3.

1.1 Types

This section covers fundamental types, including pointers, references and arrays. There are some very basic concepts covered here along with some syntactical foibles. Candidates are not expected to memorize the intricacies of the syntax but should not be confused by any unusual syntax.

Critical Information

Definition

Doing anything of practical use in C++ involves variables and functions. A variable requires a name, a type and, as a rule, a collection of meaningful actions that may be carried out on it.

Informally, types specify what set of *values* a name can contain, along with what *operations* are valid for those values. For example, an integer type allows only whole number values and has plus and subtraction operations (among others) associated with it.

All values in C++ require memory. The language itself does not specify how much memory – that is an implementation detail – but typically an `int` is represented by a single 32-bit word (4 bytes). This is about as basic as it gets:

```
int n = 42;
```

The identifier `n` is the name of a variable of type `int` and is initialized with the value of `42` (typically held in 4 bytes of storage).

Basic types

The following basic types are built into the C++ language and are known as *fundamental types*:

- `bool` – Boolean
- `char` – character
- `int` – integer
- `double` – floating-point number.

There is also `w_char_t`, the wide character type, not covered here.

The first three types, `bool`, `char` and `int`, are known as *integral* types, values that do not have a fractional part. Integral types may be promoted or cast to larger types without any loss of precision, for example from a `char` to an `int`. For assignments between integral and floating-point types (`double`), precision cannot be guaranteed; for example, the fractional part of a `double` is rounded on assignment to an `int`.

Integral types and floating-point types are called *arithmetic* types, types that allow mathematical manipulations. Arithmetic types can be mixed freely in assignments and expressions, but this practice is not recommended due to undefined behavior and potential loss of precision.

Integral types may be augmented by the keywords `signed`, `unsigned`, `short` and `long` to allow better granularity of the storage size of the type. The `unsigned` keyword, for example, frees up the complement bit, allowing larger positive numbers to be represented in the same amount of memory; `unsigned short char` typically represents an 8-bit ASCII character.

Typedef

To provide developers with a degree of flexibility both in the practicalities of naming and as a porting aid, C++ provides the `typedef` keyword to allow a type to have an alias.

```
typedef unsigned long int TUint32
```

`TUint32` is less of a mouthful than `unsigned long int` and, when porting to another system, if the underlying implementation of `long` is less than 32 bits, only the `typedef` declaration need be modified to accommodate a larger type.

It should be noted that `typedefs` are not distinct types in themselves but synonyms or aliases. There are no specific operations or type checks for `TUint32`, only for the existing type it substitutes.

Constants and enumerations

The keyword `const` provides a mechanism to reduce errors caused by inadvertently reassigning a value to an identifier. A `const` identifier has to be initialized when it is declared and cannot be reassigned any other value.

```
const int n = 42;   // n has to be initialized and cannot be reassigned
```

C++ also supports a legacy method for providing named constants, the `#define` preprocessor directive (see the information about preprocessors in Section 1.6).

```
#define N 42
```

This is a brute force approach to constants; the preprocessor is executed before the C++ compiler and replaces all the occurrences of the identifier `N` with the value `42`. The problem with this is twofold: there is no scope control and no type-checking. Using the `const` definition, `n` can only be defined once, is clearly an `int`, and has a well-defined scope. With the preprocessor directive, there is no concept of scope and `N` may be redefined at any time to contain another value of any type.

The final type of constant that C++ supports is a user-defined type enumeration:

```
enum TPriority
  {
  EPriorityIdle = -100,
  EPriorityLow = -20,
  EPriorityStandard = 0,
  EPriorityUserInput = 10,
  EPriorityHigh = 20
  };
TPriority priority = EPriorityLow;
priority = EPriorityHigh;  // (a) Fine: priority itself is not a const
priority = -10;            // (b) Error: -10 is not of type TPriority
priority = TPriority(-10); // (c) Fine: -10 is within the enumeration
                           // range and may be explicitly converted
priority = TPriority('a'); // (d) Fine! 'a' is a char - an integral
                           // type
int i = EPriorityIdle;     // (e) Fine
```

In the above example, `TPriority` is the name of a type that has a set of five constant *enumerators* with assigned values. This is not mandatory; if no values are assigned, the first enumerator value is 0 and each subsequent enumerator increments by 1.

The `enum` variable `priority` can only be assigned constant values of the `TPriority` type, although integral types can also be assigned `TPriority` values. See the comments in the example.

See Chapter 10 for information on the Symbian OS treatment of `const` values in DLLs.

Pointers

C++ provides a very simple derived type, the pointer type, which has led to some of the most powerful and, arguably, difficult to understand code ever written. For every fundamental or user-defined type there is an associated type that points to it; in pseudo-formal language, for every type `T` there is the pointer type `T*`. A variable of type `T*` simply holds the address of a variable of type `T`.

```
short int n = 42;
int* p = &n;        // The address of short int variable n
short int m = *p; // p is dereferenced and m is assigned n, i.e. 42
```

The variable `n` has the short integer value of `42`. The address of `n` is retrieved by the address-of operator, `&`, and is assigned to the pointer variable `p`. To retrieve the value of `n` from `p`, it is a matter of *dereferencing* the pointer using the dereference operator, `*`.

Note the use of the `*` operator: in a declaration statement or function parameter list, the `*` symbol signifies a pointer type whereas, in expressions, it is the dereference operator.

It is possible to have pointers to pointers or multi-level pointers.

```
// carrying on from the above example
int** pp = &p;      // pp contains the address of p which contains the
                    // address of n which contains the value 42
m = **pp;           // deref pp, deref p and assign the value of n to m
```

Thus pp contains the address of the p pointer, which in turn contains the address of the n identifier containing the value 42. Multi-layering can in theory continue indefinitely.

Null or zero

It is possible to declare an uninitialized pointer, or for a pointer to unintentionally point at some random part of memory. There is no way to determine at run-time whether a pointer contains an illegal address. To reduce this possibility, C++ has made the number zero (0) special for pointers.

A pointer containing the value 0 is not pointing at an object but is a *null pointer*. It is usual practice to declare a global constant to clarify the null value:

```
const int NULL = 0;
```

Pointer arithmetic

Pointer arithmetic is relative to the size of the type that is *pointed to*.

```
short int n;            // size of short int is 2 bytes
short int* p = &n       // p contains address of n
p = p+1;                // p has incremented by 2 bytes

int m;                  // size of int is 4 bytes
short int* q = &m;      // q contains address of m
q = q+1;                // q has incremented by 4 bytes
```

So, for short int* p, p=p+1 increments the pointer by 2 bytes, whereas for int* q, the increment is 4 bytes.

This behavior is applicable to all pointer arithmetic operations including the ++ and -- operators (see Section 1.3).

Arrays

A C++ array is a contiguous piece of memory containing individually addressable values which may be accessed using the offset operator, [ ]. In expressions, the [ ] operator takes an index to access the value at that position in the array. C++ arrays are zero-indexed.

```
int array[] = {1,2,3};  // a three-element array containing 3 integers
                        // initialized to 1, 2 and 3
int res = array[0]+array[1]+array[2];  // Addition of all the elements
```

An array behaves in the same *type-size-safe* manner as a pointer, so an array of three `short int` values requires 3*2 bytes whereas an array of three 4-byte `int` values requires 3*4 bytes. Pointers and arrays are interchangeable in some respects:

```
int* ptr = array;               // ptr contains the address of the 1st
                                // element
res=*ptr+*(ptr+1)+*(ptr+2);     // add all the elements - it can soon get
                                // confusing
```

The statement `ptr = array;` is semantically the same as `ptr = &array[0];`. This is because C++ does not support operations on entire arrays, only on individual array elements; any mention of an array name in an expression is interpreted by the compiler as a pointer to the address of its first element.

```
int aArray1[3];
int aArray2[3];
aArray1 = aArray2;  // Illegal! No array-level operations supported
```

Multidimensional arrays

In the same way as there are pointers to pointers, there can be arrays of arrays or, more accurately, multidimensional arrays.

```
int matrix[][3]={{1,2,3},{4,5,6},{7,8,9}};
int n = matrix[1][1];       // n == 5
```

The `matrix` variable is a 3 by 3 array, that is three arrays each containing three integer values. Note that in the `matrix` declaration only the first dimension may be undefined; the second and any subsequent dimension sizes must be known.

See Section 1.4 for notes and comments about argument passing to functions.

References

A reference is an alias or alternative name for an existing object. The ampersand symbol (&) indicates a reference type in a declaration. The following example shows the behavior and use of references:

```
int n = 42;         // n contains 42
int& ref = n;       // ref is an alias for n
```

```
int m = ref;        // m contains the value 42
int* p = &ref;      // semantically the address of n
int k = *p;         // dereference p to provide the reference to n, thus
                    // k==42
```

References are different from pointers in a number of ways:

- References must be initialized (when declared they must refer to an existing object) whereas pointers may be declared uninitialized.
- Once initialized, a reference cannot be changed, but a pointer may be reassigned.
- There are no direct operators that act on references, only on the value referenced. A pointer has operators that act directly on it (arithmetic and dereference).
- There is no `NULL` value for a reference.

Const pointers

Const pointers play an intricate part in C++ programming, often requiring a keen eye to ensure that the intended semantics and behavior are correct. Specifically, **a const pointer is not a pointer to const**.

```
const char* p1 = "Mouse";           // pointer to const identifier
char* const p2 = "Small";           // const pointer to non-const variable
const char* const p3 = "Horse";     // const pointer to a const identifier
p1 = "Elephant";                    // Legal to reassign non-const ptr
p2 = "Huge";                        // Illegal reassignment of const ptr
```

In the above example, `p1` is a pointer that points to the constant value `"Mouse"`, but it may be reassigned to point to another `const char` value. `p2` is a constant pointer to a non-constant value and cannot be reassigned (similar to a reference). `p3` is itself constant and also points to a constant value.

```
char* p4 = "Cat";       // pointer to non-const declaration
p1 = p4;                // pointer to const assigned a non-const variable
p4 = p1;                // Error: illegal attempt to downgrade a const
```

C++ allows a pointer to `const` to point to a non-`const` value. This initially appears counterintuitive, but what it means is the value being pointed to will be treated as a `const`. However, a pointer to a non-`const` cannot be assigned the value of a pointer to `const`, as that would remove the *constness* of the pointer to `const` type.

It is worth noting the following syntactical equivalences:

```
char const* p5 = "Duck";           // pointer to const, the same as p1
char const* const p6 = "Goat";     // const pointer to const char, the same
                                   // as p3
```

Exceptions and Notes

The `mutable` keyword, which can override `const`, is not covered here.

Exam Essentials

- Understand that C++ has a number of different types in various categories for example integral, arithmetic
- Recognize `typedef` as defining a synonym for an existing type but not a new type in itself
- Recognize enumeration types as user-defined value sets
- Specify the advantages of `const` over `#define`
- Understand the use and properties of the C++ reference type
- Specify the difference between pointers and references
- Understand the semantics of pointer arithmetic
- Recognize pointer operations and the purpose of the `NULL` pointer value
- Differentiate `const` pointers and pointers to `const`

References

[Dewhurst 2005 Items 5, 7 and 8]
[Stroustrup 2000 Chapters 2 and 4, Sections 5.1–5.6 and Sections C.1–C.7]

1.2 Statements

Critical Information

Declarations and definitions

This section broadly covers the C++ rules for declaration and definition. It does not go into too much specific detail as this is dealt with in Chapter 2.

Before an identifier or a function can be used, it has to be declared. A *declaration* does not have any execution expressions or storage requirements, and an entity may be declared any number of times throughout a program as long as the declarations are consistent. An entity *definition* contains the expressions and storage requirements for the identifier and an entity may only be defined once so that it does not have more than one possible *meaning*.

At its most basic, declaring a variable is a matter of associating an identifier with a type, for example `int n` (see Section 1.1). In this case, the declaration is also the definition as `n` requires storage (4 bytes).

Function declarations and definitions are typically separate, allowing implementation hiding and separate compilation (see Section 1.6):

```
int Add(int a, int b);      // Function declaration
// ...
int Add (int a, int b)      // Function definition
  {
  return a+b;
  }
```

The `extern` specifier, when added to the start of a declaration, tells the compiler not to allocate memory for the identifier as it is defined elsewhere:

```
extern int myInt;   // Declaration, no memory is allocated to myInt
//...
int myInt = 42;     // myInt definition
```

Scope

Once declared, an identifier has a scope or lifetime. Scope is the span of code in which an identifier is usable. In C++, scope has several levels of granularity: file or global scope, function scope, block scope, class scope and namespace scope. Class, namespace and function scope are dealt with in Chapter 2.

In file or global scope, the identifier is *visible* from its declaration to the end of the file. The `extern` keyword enables the same identifier to be declared visible in more than one file.

Block scope is the most common unit of scope. A block consists of zero or more statements enclosed by braces (or "curly brackets"), `{}`. Blocks may be nested. The following example shows two nested `if` blocks:

```
int m = 32;
if (m>0)
  {// Outer block scope
  const int n = 42;
  if (n>m)
    {// Inner block scope
    int i = n;          // outer n is in scope, i is 42
    ...
    const int n = 52; // Legal
    i = n;              // inner n is in scope, i is 52
    }
  // i and inner n are out of scope, outer n is in scope
  }// outer n is out of scope, m is still visible
```

It is important to note that the outer block identifiers are scoped within the inner block, but a declaration (and definition) of the same name in the inner block overrides the outer scope identifier until the inner block closes and the inner identifier goes out of scope.

Conditional statements

C++ supports two conditional or selection statements, `if` and `switch`, in which a condition is tested for and the result determines which action (if any) is carried out.

```
if (condition) statementOne
```

`statementOne` is only executed if the result of the `condition` expression is true (non-zero), otherwise the program flow continues.

```
if (condition) statementOne else statementTwo
```

`statementOne` is executed if the result of the `condition` expression is true (non-zero), otherwise `statementTwo` is executed.

The `condition` expression can use the following comparison operators, which return a `bool true` (or simply a non-zero value in older compilers) if the test is true, otherwise `false` (or zero): `== != < <= > >=`.

Logical operators are used to provide support for more sophisticated conditions:

- `!`, the NOT operator, negates its argument.
- ||, the OR operator, only evaluates the second argument if the first is false (or zero).
- `&&`, the AND operator, only evaluates the second argument if the first is true (or non-zero). This is a useful property when testing for validity, as in this example using pointers:

```
if (p && p->IsValid()) statementOne
```

This ensures that `p` is only dereferenced if it is non-zero, that is not `NULL`.

C++ also provides the *conditional expression*, or ternary operator `? :`, which is sometimes more convenient for simple selections:

```
char* name = isClangerShort() ? "Small" : "The Major";
// which is the equivalent of
char* name = NULL;
if (isClangerShort())
  name = "Small";
```

```
else
  name = "The Major";
```

In Symbian OS C++ programming, using the conditional expression is discouraged, as it can lead to obscure code.

```
switch (expression) statements
```

The integral value of the `expression` determines which of the `statements` is (or are) executed. The `switch` statement is well suited for testing against sets of constants and enumerations rather than purely propositional evaluations.

```
// priority is a TPriority enumeration type
// See Section 1.1 for more information
  char* statusString = "idle";
  switch (priority)
    {
    case EPriorityLow:
      statusString = "Low";
      break;
    case EPriorityStandard:
      statusString = "Standard";
      break;
    case EPriorityHigh:
      statusString = "High";
      break;
    case EPriorityUserInput: // deliberate fall through
    default:
      statusString = "User Input";
    }
```

The `priority` value is evaluated and the matching `case` is executed. If no `case` is matched, the `default` is executed. An important point to note is that unless each `case` statement is terminated with the `break` keyword, the execution will continue on to the next `case` (including `default`) regardless of the value of the `expression`.

Neither the `case` nor `default` keywords may be used outside the scope of the `switch` statement The `break` keyword is also valid in loop and conditional statements.

Iteration statements

There are three iteration or loop statements in C++: `while`, `for` and `do`. All the statements repeat a block of instructions until some condition is met or a `break` is used.

```
while (condition) statement;
```

```
for (for-init-statement; condition; expression) statement;
do statement while (condition);
```

The `while` statement is typically used in operations such as reading an input stream where there are no obvious numeric limits, making a logical test more practical.

```
Token token = start(stream); // get a token from an input stream
while (token != EOF)         // not the end of the input stream
  {
  // ... do something
  token = next(stream);      // and the next one
  }
```

But it is entirely possible to achieve the same result with a `for` loop:

```
for (token=start(stream); token!=EOF; token=next(stream))
    {
    // ... do something
    }
```

The `for` statement has an advantage as its loop variable, condition and loop variable update expression are all on one line, which reduces the chance of error, but it can make code logically unclear. For this reason, `for` loops are best suited to regular order sequences:

```
// A regular kind of loop!
for (int ii=0; ii<10; ii++)       // do something 10 times
  {
  // ... do something
  }
```

The `do` statement allows the code it is controlling to be executed once **before** the control condition is tested for. Thus, the precondition is that the loop code is safe for the first iteration. In the example below, the stream definitely contains tokens.

```
token = start(stream); // pre-condition: The stream is not empty
do
  {
  // ... do something
  token = next(stream);
  } while (token!=EOF);
```

The `break` keyword may be used to stop the flow of execution and resume it after the loop. A more finely grained manipulation of loop execution is the `continue` keyword, which stops the current iteration of the loop and starts the next iteration immediately:

```
for (token=start(stream); token!=EOF; token=next(stream))
  {
  if (token==SPACE) continue;    // ignore spaces, skip to next token
  // ... do something
  while (Param p = getParam(stream))
    {
    if (illegalParam(p) break;     // drop out
    // ... do some param processing
    } // Param p is out of scope
  }
```

The above code segment reads a stream and ignores spaces by continuing to the next token. The inner `while` loop retrieves the token parameters and tests to see if they are legal. If not, the `while` loop is terminated and drops back to the outer `for` loop. Note the declaration of the `Param p` variable in the condition expression and its scope within the inner loop only.

Exceptions and Notes

- The One-Definition Rule (ODR) is not covered in any depth here; see [Stroustrup 2000 Section 9.2.3].
- There are a number of non-conditional ways to break out of a loop: `panic`, `leave` or simply `return`.

Exam Essentials

- Know the use and properties of the declaration and definition statements
- Recognize the use of the `extern` keyword
- Cite and understand initialization of variables and their scope
- Understand the purpose, syntax and behavior of C++ loop statements (`while`, `for` and `do`)
- Specify the behavior and effect of the `continue` and `break` keywords in a loop
- Specify the syntax and behavior of C++ conditional statements (`if` and `switch`)

References

[Stroustrup 2000 Sections 6.1 and 6.3 and Exercise 6.6]

1.3 Expressions and Operators

Critical Information

Operators

Operators are usually symbols, such as `+`, `-`, `*`, but can also be names, such as `new` or `delete`, and have a number of qualities aside from their actual purpose or function:

- An operator may be unary, binary or ternary (that is may take one, two or three arguments or operands).
- Operators have associativity.
- Unary operators may be prefix or postfix.

The number of operands an operator or function takes is sometimes referred to as *arity*. All operators return a single value. The table shows some examples:

Type	Common Form*	Examples

Unary	**operator**(value)
	operator(expr)
		`n++`
		`--n`
		`!b`

Binary	value **operator** value
	value **operator** expr
	expr **operator** expr
		`a + b`
		`a >> b`
		`a && b`

*This is pseudocode notation, not formal language definition syntax

Associativity refers to the order of evaluation.

- Left-to-right associativity: `a+b+c` means `(a+b)+c` (that is, `a + b` is evaluated first)
- Right-to-left associativity: `a+b+c` means `a+(b+c)` (that is, `b + c` is evaluated first)

A great proportion of operators in C++ have left-to-right associativity.

The increment and decrement operators (`++` and `--`) have two forms: *postfix* and *prefix*.

- Prefix increment: `a=++b` means `b=b+1; a=b;`
- Postfix increment: `a=b++` means `a=b; b=b+1;`

Operator precedence

There are priority rules for the order of evaluation of operators within an expression; that is, which operator is evaluated before another. For example, the multiplication operator `*` has a higher precedence than the plus operator `+`.

`a+b*c` means `a+(b*c)`; that is, `b` is multiplied by `c` before `a` is added to the result.

This is called *operator precedence* and should not be confused with operator associativity, which is the order of evaluation of results for multiple instances of the same operator or operators of the same precedence.

In the following table, operators in the same box have the same order of precedence and the boxes are arranged with the highest precedence at the top of the table.

Operator	**Name**	**Associativity**
`::`	Scope resolution (global and class)	Left-to-right
`.`	Member selection (object)	Left-to-right
`->`	Member selection (pointer)	Left-to-right
`[]`	Subscript (array)	Left-to-right
`()`	Function call	Left-to-right
`++`	Postfix increment	Left-to-right
`--`	Postfix decrement	Left-to-right
`static_cast`	Type cast	Left-to-right
`sizeof`	Size of object or type	Right-to-left
`++`	Prefix increment	Right-to-left
`--`	Prefix decrement	Right-to-left
`~`	One's complement	Right-to-left
`!`	Logical not	Right-to-left
`-`	Unary minus	Right-to-left
`+`	Unary plus	Right-to-left
`&`	Address-of	Right-to-left
`*`	Indirection	Right-to-left
`new`	Create object	Right-to-left
`delete`	Delete object	Right-to-left
`()`	Cast (cohesion)	Right-to-left
`.*`	Pointer-to-member (objects)	Left-to-right
`->*`	Pointer-to-member (pointers)	Left-to-right

<table>
<tr><th>Operator</th><th>Name</th><th>Associativity</th></tr>
<tr><td>*
/
%</td><td>Multiplication
Division
Modulus</td><td>Left-to-right
Left-to-right
Left-to-right</td></tr>
<tr><td>+
–</td><td>Plus
Subtraction</td><td>Left-to-right
Left-to-right</td></tr>
<tr><td><<
>></td><td>Left shift
Right shift</td><td>Left-to-right
Left-to-right</td></tr>
<tr><td><
>
<=
>=</td><td>Less than
Greater than
Less than or equal to
Greater than or equal to</td><td>Left-to-right
Left-to-right
Left-to-right
Left-to-right</td></tr>
<tr><td>==
!=</td><td>Equality
Inequality</td><td>Left-to-right
Left-to-right</td></tr>
<tr><td>&</td><td>Bitwise AND</td><td>Left-to-right</td></tr>
<tr><td>^</td><td>Bitwise exclusive OR</td><td>Left-to-right</td></tr>
<tr><td>|</td><td>Bitwise inclusive OR</td><td>Left-to-right</td></tr>
<tr><td>&&</td><td>Logical AND</td><td>Left-to-right</td></tr>
<tr><td>||</td><td>Logical OR</td><td>Left-to-right</td></tr>
<tr><td>e1?e2:e3</td><td>Conditional/ternary</td><td>Right-to-left</td></tr>
<tr><td>=
*=
/=
%=
+=
–=
<<=
>>=
&=
|=
^ =</td><td>Assignment
Multiplication assignment
Division assignment
Modulus assignment
Plus assignment
Subtraction assignment
Left-shift assignment
Right-shift assignment
Bitwise AND assignment
Bitwise inclusive OR assignment
Bitwise exclusive OR assignment</td><td>Right-to-left
Right-to-left
Right-to-left
Right-to-left
Right-to-left
Right-to-left
Right-to-left
Right-to-left
Right-to-left
Right-to-left
Right-to-left</td></tr>
<tr><td>,</td><td>Comma</td><td>Left-to-right</td></tr>
</table>

Although memorization of the precedence order is not required for the ASD exam, it is important to be familiar enough to know that scope resolution, member selection, pointer operators and unary operators are towards the top, with arithmetic and conditional operators making up the middle and the lowest being the assignment operators. Note the low precedence of the logical AND and OR.

Exceptions and Notes

`a++ + a++` is undefined: the variable is read twice in the same expression, producing ambiguities during the evaluation.

Exam Essentials

- Specify the syntax and meaning of unary and binary operator expressions
- Understand the difference between precedence and associativity
- Recognize common operator categories including logical, prefix and postfix
- Demonstrate awareness of general operator precedence and associativity rules

Reference

[Stroustrup 2000 Sections 6.2.1–6.2.5, Sections 11.1 and 11.2, Section A.5]

1.4 Functions

This section deals with the syntax and semantics of non-member functions, that is functions that do not belong to a class.

Critical Information

Declaration

A function may be declared separately from its definition and declared more than once, provided the declarations are exactly the same. A function declaration is typically referred to as a function *prototype*. The clear separation between declaration and prototype enables a function to be used before it is defined.

For the compiler to ensure that a function is called legally, the function prototype needs to contain the function name, a list of parameter types and a return type. The parameter names are not required in the declaration.

Parameter lists in themselves are declarations containing the type name and an optional qualifier (`const`). As well as fundamental and user-defined types, pointers (`*`), references (`&`) and arrays (`[]`) can be passed.

A parameter may be initialized with a *default* value, so that when the function is called without a value for this parameter, the default value is assumed. The default arguments must always be at the end of the parameter list and can only ever be declared once, regardless of the number of times the function is declared.

```
// an illustrative example
enum OperatorType {ADD, SUBTRACT /* , ... etc */ };    // Enum for
                                                       // selecting
// Declaration
int math (const int a, const int b, const OperatorType optype = ADD);
                                                        // default add
// Definition - note the default value cannot be declared again
int math(const int a, const int b, const OperatorType optype /*=ADD*/)
  {
  switch (optype) { /* return something... */ }       // the mechanics
  }
// In use
int res = math(2,2);                           // res == 4
res = math(2,2,ADD);                           // res == 4
res = math(math(2,2,SUBTRACT),2);              // res == 2
```

The `math` function takes two operands and a selection `enum`, which has the default value of ADD (not re-declared in the definition), and returns an `int`.

Inline functions and macros

For simple and often-used functions, a C developer would use the `#define` macro directive to remove the overhead associated with function calls and improve efficiency.

```
#define ADD(a,b) (a+b)
int res = ADD(2,2);  // whenever ADD is encountered, it is substituted
                     // with (a+b)
```

Macros are generally discouraged in C++ due to lack of type checking and behavioral problems when the `#define` is more complex.

Matching of `if-else` statements does not behave as expected:

```
// the if problem
#define TEST(A) if(cond) A;
int x = 0;
if (okay) TEST(x) else x = 42;    // The if in the macro is matched to
                         // the else rather than the intended if(okay)
```

If a `#define` contains a series or block of statements, there are unexpected results in loops:

```
// block problem
#define BLOCK stmt1; stmt2; stmt3;
for (ii=0;ii<10;ii++) BLOCK; // Only stmt1 is executed inside the loop
                             // stmt2 and stmt3 are executed outside
```

Pseudo-macro parameters do not behave as expected:

```
// Multiple substitution
#define SQUARE(n) (n*n);
int n=2;
int res = SQUARE(n++);
// n is now 4 i.e. incremented twice
```

The `inline` keyword provides C++ developers with the same efficiency as a macro without the behavioral ramifications. The function body is substituted wherever the call is made.

```
inline int square (int n)
  {
  return n*n;
  }
// ...
int n = 2;
int res = square(n++);
// n is 3 i.e. incremented once, the desired effect
```

The `inline` keyword is only a request to the compiler. It does not guarantee the substitution and it may still be treated as a normal function call. When a member function is fully defined in a class header file, in effect it becomes inline by default.

Passing an unspecified number of parameters

To recap, a function prototype has a return type, a name and a parameter list. The parameter list contains the parameters that the function uses, and parameters at the end of the list may be initialized with a default value.

A function can also allow an *unspecified* number of arguments to be passed to it. The following example shows the syntax for declaring and using such a function, and how to access the arguments within the function body.

```
enum OperatorType {ADD, SUBTRACT};
// Declaration
int math (const OperatorType ty ...); // the ellipsis ... means any number
```

```
// Definition
int math (const OperatorType ty ...)
  {
  va_list ap;              // param list
  va_start(ap,ty);         // where to start in the list i.e. after ty
  int x = va_arg(ap,int); // go through the parameters, which are ints
  int res = 0;
  while (0!=x)             // while not zero
    {
    switch (ty)
      {
      case ADD:            // other cases, e.g. SUBTRACT, not used
      default:
        res+=x;
      }
    x = va_arg(ap,int);   // Get next param
    }
  va_end(ap);              // clean up the stack frame
  return res;
  }
// Called
int r = math(ADD,1,2,3,4,5);
```

`e32def.h` contains the `va_list`, `va_start` and `va_end` macros.

The ellipsis (`...`) tells the compiler that the number of arguments that follow the `const OperatorType` is unknown. Variable argument macros are provided to access the list:

- `va_list` is a pointer to the list of arguments.
- `va_start` tells the code where to start the list.
- `va_arg` gets the next parameter, supplying its type.
- `va_end` cleans up the stack.

Passing arguments by value and by reference

C++ provides two ways of passing arguments, by value and by reference. When an object is passed by value, a local copy is made on the function local stack and all subsequent modifications are made to that copy without any side-effect modification to the original copy. When a reference is passed, no local copy is made and all manipulations are carried out directly on the original object.

As passing by value actually copies the object, large values such as data structures can cause efficiency problems not only in taking up local stack space but in the action of copying. Passing by reference only copies the address, typically 4 bytes, and is therefore more efficient.

To prevent any unnecessary modifications to the referenced object, the `const` keyword is used to indicate that it will not be modified and is being passed as a reference for efficiency reasons.

```
void function (int val,                 // by value
               int& ref,                // by modifiable reference
               const int& ReadOnlyRef, // by non-modifiable reference
               int* ptrVal)             // by value
  {
  val = 42; // val will only contain 42 for the scope of this function
  ref = 42; // The variable referenced is now 42
  ReadOnlyRef = 42; // *** Illegal *** attempt to modify a constant
  *ptrVal = 42;
  }
// In use
int a = 0,b = 0,c = 0,d = 0;
function (a,b,c,&d); // post-conditions: a=0,b=42,c=0,d=42
```

There can be some confusion as to how the ampersand operator, `&`, is used in declarations and expressions (see Section 1.2). In declarations of variables or functions, `&` signifies a reference type, whereas in statements and expressions, such as when calling a function, `&` signifies address-of.

In the example above, when `function` is called, the address of `d` is passed (parameter `ptrVal`). Incidentally there is no concept of "pass by pointer"; a pointer is an address and the address is passed by value.

Passing arrays to functions

Passing arrays as arguments is a special case. Consider the following examples:

```
void foo (int array[])              // pass an array
  {
  array[0] = 42;                    // assign 42 to the first element
  }
void bar (int* p)                   // pass a pointer to an int
  {
  p[0] = 42;                        // assign 42 to the first element!?
  }
// Calling
int myArray[] = {1,2,3};            // an initialized 3 element array
foo(myArray);                       // pass the array
bar(myArray);                       // pass the array - the same way
```

In the above example `foo(int array[])` has the expected syntax for passing an array as argument. What in fact happens semantically is a little less obvious: as an array variable is always interpreted as a pointer to its first element (see Section 1.1), the parameter `int array[]` is converted to `int* array` when passed as function parameter. This is sometimes called array-to-pointer *decay*.

Thus `foo()` and `bar()` are semantically the same and interchangeable. An implication of array-to-pointer decay is that an array can never be passed by value. There is no local copy of `array[]`; in `foo()`, any modifications to `array` directly change `myArray`.

The size of the array is not known to the function. To prevent out-of-range subscribes, that is, accesses to locations past the end of `array`, it is good practice either to pass the length as an additional parameter or to define a *termination* value for the last element.

The same rules for multidimensional array declarations apply to passing parameters: only the first dimension may be undefined. The size of the second and any subsequent dimensions must be known, or "bound", before the assignment of values.

```
void foobar (int array[][3])
  {
  array[1][1] = 42;
  }
int array[][3]= {{1,2,3},{1,2,3},{1,2,3}};
foobar(array);
```

The syntax for interchanging pointer notation with multidimensional arrays is tricky. `barfoo (int* array[3])` takes a single array of `int` pointers, whereas `barfoo (int (*array)[3])` takes the same effective parameter as `foobar (int array[][3])`, a multidimensional array of `ints`.

Return values

A value must be returned from any function that is not declared `void` and, conversely, `void` functions cannot return a value. Only references and values (including pointers) may be returned. A function should never return a reference or pointer to a variable local to the function scope, that is on the stack.

```
// bad pointer return
int* foo ()     // returns an int pointer
  {
  int n = 42;
  return &n;  // returns the address of n
  }           // n is out of scope, thus the return address is invalid
// bad reference return
int& bar ()     // return a reference to an int
  {
  int n=42;
  return n;   // the compiler creates a reference to n
  }           // n is out of scope, thus the reference is now invalid
```

The following code is an example of returning a reference safely, because it refers to a member variable which does not go out of scope:

```
int& TTemperature:GetTemperature() {return iTemperature;}
```

Particular attention is required when returning modifiable references, because the object can then be modified by the calling code.

```
TTemperature temperatureGauge;
int& temp = temperatureGauge.GetTemperature();
++temp;          // the member variable TTemperature::iTemperature has
                 // now been modified
```

Pointer to function

In the same way as pointers to variables, it is also possible to have pointers to functions. C++ allows the address of a function to be acquired in the same way as the address of a variable, by using the address-of operator, `&`.

```
// simple function definition
void foo(int& n)
  {
  n = 42;
  }
// Function pointers
void (*bar)(int& n) = &foo;    // the identifier bar is a pointer to a
                  // function whose prototype must contain the parameter
                  // list of a single ref to an int
                  // It is initialized with the address of foo()
int n = 0;
bar(n);           // Call foo via bar
```

As function pointer declarations are very similar to actual functions, there are some syntax specifics that require careful attention. A pointer-to-function identifier must be put in brackets `(*bar)`; this tells the compiler it is parsing a pointer to a function and not another function declaration or a data-pointer declaration. The return type (`void`) and the parameter list (`int& n`) indicate the prototype of the function the pointer will point to, that is `(*bar)` points to a function with the following prototype:

```
void function-name (int& n);
```

For brevity of code, the use of the address-of operator, `&`, and the dereference operator,`*`, is optional.

```
void (*bar)(int& n) = foo;
bar(n);
// is the same as
void (*bar)(int& n) = &foo;
(*bar)(n);  // braces required to indicate not to dereference a return
            // value
```

When passing parameters, function pointers behave in same way as other pointers.

```
// foo definition
void foo(int& n)
  {
  n = 42;
  }
// foobar definition with function pointer
void foobar (int& n, void(*callback)(int&n))
  {
  callback(n);
  }
// in use
int n = 0;
foobar(n,foo);    // n is now 42
```

Function pointer arguments are typically used in C as a method of dynamically selecting a function without knowing its name at compile time, thus enabling modifications to program behavior without unnecessary recompilation. For example, `foo()` could be completely rewritten without any modification to `foobar()`. In C++, virtual functions have superseded the need for function pointers, but it is still important to recognize the function pointer syntax. (For more on virtual functions, see Chapter 3.)

Exam Essentials

- Understand the syntax of a function prototype
- Cite the purpose of the `inline` keyword
- Understand the rules for passing default arguments and an unspecified number of arguments to a function
- Recognize value, reference, array and pointer parameter passing
- Understand the scope of function blocks and return by reference and value
- Specify the syntax for pointers to function assignments
- Recognize pointer to functions as callback parameter arguments

References

[Dewhurst 2005 Items 6, 14 and 25]
[Meyers 2005 Items 20, 21 and 28]
[Stroustrup 2000 Chapter 7 and Section C7.3]

[Sutter 1999 Item 43]
[Sutter and Alexandrescu 2004 Item 98]

1.5 Dynamic Memory Allocation

There are three types of storage scope: *static*, *automatic* and *dynamic*. Scope is defined by the minimum duration or lifetime of an object. Static and automatic storage occur when the objects are defined (see Section 1.2). Static objects last for the duration of the program and are indicated by the keyword `static`. Automatic objects only exist within the block where they are defined. The reserved word `auto` is now moribund in C++. This section describes dynamic storage allocation.

Critical Information

New and delete operators

Heap memory, or free storage, allocation and deallocation are provided by the operators `new` and `delete`. The `new` operator takes one argument, the size of the object type, and allocates the amount of memory required by that type, returning the location of the memory allocated.

```
int* p = new int;     // allocate 4 bytes == sizeof(int);
char* pc = new char; // allocates 1 byte == sizeof(char);
//... p and pc are in scope until ...
delete p;             // destroys the memory - frees 4 bytes
delete pc;            // ... frees 1 byte
```

The duration of storage for `p` and `pc` is until `delete` is called to destroy them, thus they will exist outside the block they were defined in.

```
int* foo ()
  {
  int* p = new int; // p variable contains the address of allocated heap
  return p;         // returns by value.
  }                 // p out of scope but the allocation is 'alive'
int* p = foo();       // address is now contained by p i.e. outside foo's
                      // function scope.
// do something ...
delete p;           // now the memory allocated by foo is freed
```

For arrays, C++ provides an additional pair of operators, `new[]` and `delete[]`. Along with the type size, the number of elements to be allocated is also passed:

```
int* p = new int[10];    // 10*sizeof(int)
```

```
p[0] = 42;          // or use *p notation
delete []p;         // note no element size value
```

It is important that `delete []` (array delete) is called. Calling the `delete` operator only frees the first element and in doing so orphans the other nine elements (9*4=36 bytes memory leaked until the program ends).

Placement new

At its most basic, the *placement* `new` operator tells the compiler not to use the default `new` operator but another *implementation* or overload of `new`. If another overload is not explicitly declared by the developer, placement `new` is used to construct objects at a specific address passed to it – that is, either pre-allocated memory or memory-mapped hardware.

```
unsigned char* heap = new unsigned char[1000];   // heap allocate
int* p = new (heap)int;                          // placement new
```

There is no equivalent *placement delete*, thus it is left to the developer to ensure that `p` is not deleted as no new memory was actually allocated. The `heap` memory must eventually be freed. Placement `new`, used like this, is limited – in the example below the `q` identifier points to exactly the same memory as `p`.

```
int *q = new (heap)int; // q is now pointing at exactly the same
                        // memory location as p
// *** better
int* q = new (heap+sizeof(int))int;
```

This is messy; arguably a better approach would be to deal with the arithmetic inside the `new` operator itself. C++ allows developers to rewrite the global `new` operator, `void* operator new (size_t size)`, but, paraphrasing Stroustrup, "this would not be a task for the fainthearted."

Symbian OS overloads the global operator `new` to take a `TLeave` parameter, in addition to the implicit size parameter. The `TLeave` parameter is ignored by operator `new`, since it is only used to differentiate it. The implementation calls a version of the Symbian heap allocation function that causes a leave if there is insufficient memory for the allocation. See Chapter 5 for more information about Symbian OS leaves.

Exam Essentials

- Understand C++ free store allocation scope using the `new` and `delete` operators
- Recognize the syntax and purpose of placement `new`

References

[Dewhurst 2005 Items 34–7]
[Meyers 2005 Chapter 8]
[Stroustrup 2000 Sections 10.4.5–10.4.11, Section 19.4, Section C.9]
[Sutter 1999 Items 35 and 36]
[Sutter and Alexandrescu 2004 Items 45 and 46]

1.6 Tool Chain Basics

This section deals very broadly with C++ tools and file organization, and serves as an introduction to the more detailed discussion of the Symbian OS tool chain in Chapter 14.

Critical Information

File and program structure

The smallest unit of compilation is a file. If only one character changes in that file then the whole file needs to be compiled again. Well-written programs are typically organized into a number of files containing source code that is organized in logical groups – usually based on the design architecture.

A C++ compiler has two distinct stages when it is passed a source file to build: macro preprocessing and compilation into object code.

Preprocessor

The preprocessor parses through the source file scanning for the directives. A directive starts with the hash sign, `#`.

Directive	Description
`#include`	Includes another file, usually header files `*.h`
`#define`	Allows the user to define a macro. `e32def.h` contains the most commonly used Symbian OS macros.

`#ifdef`
`...`
`#endif`
	May be used to turn code on and off, typically in conditional compilation

The preprocessor will expand all the macros in the code. This also entails expanding the `#include` directive which adds the specified file to the current source file (see Section 1.4).

Separate compilation

The C++ language already provides the mechanism to separate variable and function declarations from definitions. To extend this feature and allow separate compilation, the `#include` directive allows the placing of declarations into separate files (`*.h` files) from their definitions (`*.cpp`). Thus, a third party may include the header and compile against it without the accompanying source definition file.

Any function definitions contained within a header file are automatically treated as `inline`, since the whole header file, including definitions, is effectively inserted into the including source file. Thus, if a definition is modified, all files that include it must be recompiled; otherwise different files may end up using different versions of the same function.

As function definitions – the code and algorithms – are more likely to change than their prototype declarations, it is recommended that `inline` functions be kept to a minimum and used only for very simple common functions (see Chapter 16 for more information).

Compiling and linking

After preprocessing, the compiler stage is executed. This is separated out into lexical analysis, syntax analysis, symbol-table generation and code generation. For the most part the source file is treated as a stream which is tokenized; a recursive syntax tree is built and validated and the target code is generated.

The compiler may be instructed to generate additional symbolic data binding the object code to the source code and associated symbols. This is used by a debugger to halt and step through the execution of the program.

Using `#ifdef _DEBUG` (typically) the developer may add debugging-specific information (for example log files) and checks including pre-condition or value assertions, for example:

```
ASSERT(p!=NULL);
```

More sophisticated on-target debugging is also possible where code is executed on a native target device, usually a development board. Using a *debugger stub* on the target, commands and data may be sent back and forth to a PC via a communication port. An example is MetroTRK for Symbian OS.

Once the target code has been generated along with resources such as UI components (see Chapter 14), the object code is linked with the other compiled files in the program, the C++ libraries and any third-party libraries, including any extensions specific to an operating system (for example, S60).

At this stage, any inconsistency in declarations and definitions will become apparent, usually in the form of an unresolved external. All `extern` declarations are resolved during linkage. Finally, on successful linkage, the program is ready for execution.

Exam Essentials

- Understand the function of the C++ tool chain (for example compiler and linker)
- Recognize the lexical and syntax-parsing stages of compilation
- Be able to describe the purpose of the C++ preprocessor, specifying common directives
- Understand the role `inline` functions play in C++
- Know how to use the `extern` keyword

References

[Stroustrup 2000 Chapter 9 and Section 24.4]

2

Classes and Objects

Introduction

An in-depth discussion of object-oriented (OO) design and programming would be out of scope for an exam primer. An Accredited Symbian Developer (ASD) is only expected to know the core principles of OO design and programming which, in truth, are only fully understood after a number of years' experience.

This chapter covers the C++ class and object aspects of OO that an ASD candidate should know and understand. There is some basic material here, in part because it may be in the exam, and in part to provide context for the more advanced subjects. As with the other chapters, the reference section should be used to verify and validate understanding.

2.1 Scope and C++ Object-Oriented Programming (OOP) Support

This section covers the groundwork for C++ classes. It recaps some of the scope rules covered in Chapter 1 and introduces the syntax for class declarations and definitions.

Critical Information

Scope, namespaces and logical groupings

As introduced in Section 1.2, an identifier has scope or visibility to other identifiers and expressions. C++ has introduced some sophisticated language mechanisms for extending the scope rules inherited from C to support object-oriented principles of data abstraction, hiding and encapsulation.

Consider the following problem of two functions with the same name:

```
// Symbian_MusicInterface file
void MusicPlayer(){}              // Definition

// S60_MusicInterface file
void MusicPlayer(){}              // Definition
```

There are two `MusicPlayer()` function definitions in two separate library files linked to the same program. As the function prototypes are identical, the One-Definition Rule is broken – the same function is defined twice (see Section 1.2). C++ provides the `namespace` mechanism to provide distinction between the two functions:

```
// Symbian_MusicInterface file
namespace SymbianOS
  {
  void MusicPlayer (){}
  }

// S60_MusicInterface file
namespace S60
  {
  void MusicPlayer (){}
  }

// In use
if (music==Sibelius)
  S60::MusicPlayer();
else
  SymbianOS::MusicPlayer();
```

In essence the `namespace` keyword provides a method for binding an identifier to the block containing it. In this example, the `MusicPlayer()` function definition is bound separately to `SymbianOS` and `S60`. The `namespace` identifier, used in conjunction with the scope resolution operator `::`, provides access to the correct `MusicPlayer()` function.

Namespaces also provide rudimentary encapsulation, or logical grouping. They are said to be "open"; that is, repeated declarations of the same namespace allow additional names to be added.

```
// In the header file
namespace SymbianOS_MusicPlayer
  {
  void Rewind();
  void Play ();
  }

// In the implementation
namespace SymbianOS_MusicPlayer     // namespaces are "open"
  {         // subsequent declarations allow more names to be added
```

```
  int iCurPos=0;            // Added
  int iVolume=11;           // Added

  void Rewind()
    {
    iCurPos=0;
    }

  void Play()
    {
    ++iCurPos;
    // ... Omitted for clarity
    }
  }

// In use
SymbianOS_MusicPlayer::Play();
SymbianOS_MusicPlayer::Rewind();

// or
using namespace SymbianOS_MusicPlayer; // using directive for clarity
Play();
Rewind();
```

Specific design issues aside, the above example allows the `Rewind()` and `Play()` functions to be encapsulated or grouped together in one logical namespace. The implementation details are hidden in a separate file and the user sees only the essential functions required to use the music player. In other words, the data is hidden and the core functionality is abstracted. The `using` keyword removes the need to use the scope operator on every function call.

Properties of classes and objects

A *class* is a user-defined type. It provides a mechanism for cohesive grouping of data and functions. It describes the interface of the type and defines the implementation.

An *object* is an *instantiation* of a class, that is, a dynamic instance that has its own identifier, scope and storage. More formally, an object is said to have:

- State – an object is dynamic and changes over time
- Behavior – defined by the class but executed by the object
- Identity – a class may have more than one object, with each object instance having a unique name.

In its purest form, a class is an interface: it has no public data, only internal state variables, constants and storage variables. Consider the following example (the syntax will be explained in more detail in subsequent sections):

```
// Foo declaration and definition
// This is a contrived example purely to illustrate state and data hiding
class Foo       // Foo is a user-defined type of numbers in the range 1..10
  {
public:
  // Interface Member Methods
  Foo(int n = 1)                  // constructor
    {
    if (InRange(n))               // only allow values 1..10
      {
      iData=n;                    // store value
      iValid=true;                // initialized
      }
    else
      iValid=false;               // not a valid value
    }

  void operator=(int n)           // Assignment
    {
    if (InRange(n))               // only allow values 1..10
      {
      iData=n;                    // store value
      iValid=true;                // initialized
      }
    else
      {
      iValid=false;               // not a valid value
      // the developer may wish to set iData to a null value
      }
    }
private:
  // Member data
  int iData;               // internal storage member
  // State
  bool iValid;             // is the object initialized i.e. valid?
  // Helper methods
  bool InRange(int n) {return n > 0 && n <= 10; }
  };
// in use
Foo n = 5; // n is an object instance of the Foo class

n = 8;  // valid
n = -2; // error! Object now in invalid state
n = 42; // error! Object now in invalid state
```

In this example, `Foo` is the class representation of a type that contains only whole numbers in the range 1 to 10. It has a `public` interface consisting of a *constructor* used for the initialization of an object and an *assignment operator* = to allow values to be assigned to an object. A `private` section, hidden from users of the object, contains the implementation and internal representation of the data and its state.

It is important to note that `n` is of type `Foo` and not `int`. The class `Foo` has a constraint (only values between 1 and 10) which `int` does not, and therefore a different integral value, for example 42, may be a valid `int` but not a valid `Foo`.

The constructor and assignment operator allow user-defined types to be used in the same manner as built-in types. Semantically, however, it is left to the developer to ensure the assignment operator is only called when it's appropriate to assign a value.

Data storage

As stated in the preceding section, the class statement is used to group functions and data together. C++ also provides the `struct` and `union` statements, which originate from C and share the same basic form.

A `struct` is exactly the same as a `class` with the exception that `class` members are `private` by default and `struct` members are `public` by default.

A `union` is a little different:

- A `union` can only hold the value for one member at a time and all members are allocated the same address space
- Storage is allocated for the largest data member
- Data members cannot be objects with user-defined constructors, destructors or copy operations
- All members are public by default.

A `union` may be anonymous or named.

```
union Bar                  // All members are public
  {
  char ch;
  int  n;                  // Memory is allocated for the larger member
  };

struct Foo                 // A struct with all members public
  {
  int m;                   // struct data member
  Bar bar;                 // member of a union Bar (no pun intended)
  union                    // anonymous
    {
    char ch;               // members are accessed directly, see below
    int  n;
    };
  };

// In use
Foo foo;                   // Declare and define a Foo struct object
foo.bar.ch='a';            // assign a value to the named union data member
foo.ch='a';                // assign a value to the anonymous member
```

This example also introduces the *member select operator* `.` which is used to access members of `class`, `struct` and `union` objects.

Exceptions and Notes

- Classes are structural devices whereas types are concepts. In C++, unlike in some other OO languages, types and classes are treated in the same way. A type in C++ is represented as a class with public interface functions.
- A singleton class is a class which can only have a single object instance with file or global scope (see Chapter 1).

Exam Essentials

- Understand the scope and lifetime properties of blocks and namespaces
- Understand C++ support for data abstraction
- Specify the attributes of a C++ object with respect to object-oriented programming
- Know the syntax of a class declaration
- Cite the differences between a class and an object
- Differentiate between basic data structures (`struct` keyword) and classes

References

[Ambler 2001 Section 8.2 (note that the discussion concerns Java rather than C++)]
[Gamma *et al.* 1995 Chapter 3]
[Stroustrup 2000 Section 2.5, Sections 10.2.8 and 10.4, Section C.8]

2.2 Constructors and Destructors

This short section covers object construction and deletion before moving on to the broader topic of class members.

Critical Information

Constructors

A constructor is a member function that is automatically called when a class object is created. Its purpose is to initialize the state of the object. Constructors have the same name as the class itself and are usually public.

Constructors can take arguments and may be overloaded, that is, more than one constructor may be defined with different parameter lists (see Section 3.3). A constructor that has no arguments, or only arguments with default values, is called a default constructor.

A constructor cannot be static; it belongs to an object and has a `this` pointer (see Section 2.3). A constructor can neither have a return value nor be virtual. It is not possible to take the address of a constructor.

```
class Bar
  {
public:
  Bar(){}                // automatically called by Foo, below
  ~Bar(){}               // see destructors section
  };

class Foo
  {
public:
  Foo(){iVal=0;}         // Default constructor
  Foo(int n){iVal=n;}    // Overloaded conversion constructor
  ~Foo(){}               // see destructors section, below
private:
  int iVal;
  Bar iBar;              // Class member object
  };

// In use
Foo foo;                 // Default constructor is called
Foo bar(42);             // Overloaded conversion constructor is called
Foo boo = 42;            // Overloaded conversion constructor is called
```

Constructors for member objects are called implicitly. This means that `iBar.Bar()` is automatically called when the `Foo` constructor is invoked (before the body of `Foo()` is executed).

The overloaded constructor takes one argument, of a different type to the class itself, making it a *conversion constructor*. Thus `Foo boo = 42;` is a valid initialization, as the compiler will pattern match `42` to the `Foo(int n)` constructor. This is a syntax nicety that can lead to unnecessary conversions in more complex code.

Declaring the constructor `explicit` prevents the constructor being called implicitly. In this case `Foo boo = 42;` would be flagged by the compiler as an illegal type conversion.

```
explicit Foo(int n);
```

The next example shows the use of the constructor *initialization list*. C++ provides this as a means for constructors to ensure that class constants and references are initialized. Note the use of an integral type constructor `iConstVal(42)`. All built-in types have constructors.

```
class Foo
  {
public:
  Foo(Bar bar):iRef(bar),iConstVal(42)  // Const & ref value initialized
    {//body ...}
private:
  const int iConstVal;      // const must be initialized
  Bar& iRef;                // references must be initialized
  };
```

The initialization list is the only way to initialize references and constants in a class; attempting to assign values in the body of the constructor (for example `Foo(Bar){iConstVal=42;}`) would be an illegal use of the assignment operator on a `const` or reference member. Leaving them uninitialized would also result in a compiler error, as constants and references have to be initialized.

Member objects are initialized in the order they are declared in the class, not the order specified in the constructor initialization list.

Destructors

A destructor, like a constructor, has the same name as its class but the destructor's name is preceded by a tilde, for example `~Foo()`. A destructor is called when an object is destroyed, either by going out of scope or by calling `delete` on a pointer to an object.

A destructor automatically calls the destructor for each of its member objects, for example the member destructor `iBar.~Bar()` is implicitly called by `~Foo()` in the example above. The member destructors are called after the body of the class destructor, in the reverse order of their declaration in the class.

Destructors are typically used to release allocated memory owned by the object. They do not take any arguments and cannot be overloaded. Destructors may be virtual to ensure correct destruction order in a derived class hierarchy (see Chapter 3).

Copy and assignment operators

To ensure user-defined types provide the same behavior as built-in types, C++ provides the concept of *copy constructors* and *assignment operators*. Their purpose is to allow class objects to be initialized with objects of the same type and to assign objects to other objects (of the same class).

```
class Foo
  {
public:
  // Construction and destruction
  Foo() {iVal = 0;}
```

```
  ~Foo(){}

  // Copy constructor
  Foo(const Foo& foo)
    {
    iVal = foo.iVal;
    }

  // Assignment operator
  Foo& operator=(const Foo& foo)
    {
    if (this!=&foo)       // check to see it's not assigning itself
      iVal = foo.iVal;
    return *this;
    }

private:
  int iVal;
  };

// In use
Foo foo;                 // normal constructor
Foo boo(foo);            // copy constructor
Foo fooboo = boo;        // also a copy constructor
boo = foo;               // an assignment
```

Copy construction is straightforward: the constructor is passed a reference to the object to be copied, of the same type as itself. The reference is declared `const` to ensure the referenced object is not modified. All initializations call the copy constructor; the statement `Foo fooboo = boo;` may appear to be an assignment, but it is an initialization of the object being declared.

The final expression in the example is an assignment. The assignment operator uses the `this` pointer to ensure it isn't being assigned to itself, and returns itself by reference. Note that, if the assignment operator returned a value rather than a reference (`Foo` instead of `Foo&`), the copy constructor would be invoked to create a new `Foo` object (and the destructor to destroy the original `this`).

It is good practice to ensure the behavior of copy constructors and assignment operators is the same.

What the compiler does for you

The compiler automatically creates a default constructor, a copy constructor, an assignment operator and a destructor if they are not already declared. The constructor and destructor call the constructors and destructors, respectively, for all member variables and any base classes. The copy constructor and assignment operator copy the member variables using their respective copy constructors and assignment operators.

Problems can arise when dealing with pointers and references. A pointer located in a source object `foo` will have only its current value assigned or copied to a target object `bar`. Both objects will point to the same referenced variable, and if `foo` deletes its member pointer it will leave `bar`'s member pointing to an undefined part of memory. (Note that default destructors do not delete pointers.) Similar problems occur with copying references, and a reference, once initialized, cannot be reassigned.

Compiler implementations vary in dealing with this problem. But as a general rule, for any class that allocates memory to its pointer members or has non-trivial references, the class designer must provide a copy constructor, an assignment operator and a destructor to ensure the expected behavior.

Exam Essentials

- Understand the order of construction and destruction for the class and its member variables
- Recognize implicit constructor invocation (overloading and pattern matching) and the role the `explicit` keyword plays in constructor declaration
- Specify what the compiler automatically generates for user-defined classes
- Understand the purpose and use of copy constructors (including parameter passing)
- Understand the difference between assignment and initialization
- Specify the required function members needed in a class to support safe ownership of pointer data

References

[Dewhurst 2005 Items 13, 32 and 41]
[Meyers 2005 Chapter 1 (Item 4) and Chapter 2]
[Stroustrup 2000 Section 10.2 and Section 11.7]
[Sutter and Alexandrescu 2004 Items 47–50]

2.3 Class Members

Section 2.1 introduces the `class` statement as the C++ mechanism for user-defined types. Class members include variables, constants and functions, the last more accurately referred to as methods when defining class behavior. A class may also contain classes and have `friend` relationships with other classes.

This section takes a closer look at access rules and member declarations and introduces the `this` pointer.

Critical Information

Access rules

Class access control is a straightforward concept that specifies three access permissions for using class members: `public`, `protected` and `private`. A user of a class is concerned only with the `public` interface and has no access to either `private` or `protected` class members. The `private` specifier hides members from class users and derived classes, whereas `protected` hides members from class users but allows them to be accessed by derived classes (see Chapter 3 for more on inheritance). Internally, all members have access to each other.

Friends

Access control is further *extended* by the `friend` specifier; a friend is a relationship rather than an access control statement. The `friend` specifier allows non-member functions to access `protected` and `private` members of the host class. Individual functions or entire classes may be declared as friends.

The code example demonstrates non-member `friend` functions, `friend` classes and also introduces method overloading.

```
// Forward declaration of Foo so it may be made a friend of Bar
class Foo;

class Bar
  {
public:
  void Set(int n) {iVal=n;}      // normal modifier
  friend class Foo;              // Foo is declared a friend of Bar
  // Non-member operator function is declared a friend
  friend Bar operator+(const Foo&, const Bar&);
private:
  int iVal;
  };

// Foo class - friend class of Bar
class Foo
  {
public:
  void Set(int n) {iVal=n;}                 // normal modifier
  void Set(Bar& bar){iVal=bar.iVal;}        // overloaded modifier
  // Non-member operator function is declared a friend
  friend Bar operator+(const Foo&,const Bar&);
private:
  int iVal;
  };
```

```
// The addition operator which is a friend of both Foo and Bar and
// thus has access to private members in both.
Bar operator+(const Foo& foo, const Bar& bar)
  {
  Bar res;                          // Local variable
  res.iVal = foo.iVal+bar.iVal;       // Access to private members
  return res;                       // returns value note
  }

// In use
Bar bar;
Foo foo;
bar.Set(42);                    // bar's iVal contains 42.
foo.Set(bar);                   // foo's iVal now contains 42.
Bar foobar = foo + bar;         // foobar's iVal now contains 84 (42+42)
                                // Bar default copy constructor is invoked
```

The `friend class Foo;` statement means that class `Foo` has access to all the members in `Bar`. This is a unidirectional declaration, as `Bar` does **not** have access to the `private` and `protected` members in `Foo`.

In the `Foo` class method `void Set(Bar& bar)`, the private `Bar` member `bar.iVal` is visible and may be manipulated. `Foo` cannot **directly** access `Bar` members as if they were in its own class scope because `Foo` has only been granted access to hidden members of `Bar` objects passed to it.

Overloading is a loose form of ad-hoc polymorphism where multiple member functions have the same name but are distinguished by their parameter types, such as `Set(int n)` and `Set(Bar &bar)`. See Chapter 3 for further information about polymorphism.

The second type of `friend` usage in the above example is the declaration of non-member `friend` functions. `Foo` and `Bar` both declare the operator + as `friend`, allowing it access to the internal `iVal` member variable in both classes. The value is created on the stack, but the statement `Bar foobar = foo + bar;` invokes the copy constructor to copy the function return value. It is usual to provide a matching operator that returns a `Foo` as well.

Note that `friend` relationships should be declared only between classes which are semantically very close, and only when all other design options are exhausted. A `friend` should never be declared purely to allow access to private members.

Nested class

Class members include data and state variables, constants, references and methods (functions and operators). It is also possible to have member classes. A member class is not a pointer to a class or an instance of a class, but a member that is a class declaration. Classes declared within the scope of another class are called *nested classes.*

```
class Outer
  {
public:
  int iOuterVal;

  class Inner
    {
  public:
    int iInnerVal;
  private:
    int iInnerPrivateVal;
    };
private:
  int iOuterPrivateVal;
  };

// In use
Outer Foo;                  // no Inner object is created
Outer::Inner Bar;           // no Outer object is created
int n = Foo.iInnerVal;      // illegal!
n = Bar.iOuterVal;          // illegal!
```

A nested class is purely a class that is declared within the scope of another class. When an object of class `Outer` is created, the `Inner` class is not instantiated, and vice versa. An `Outer` object does not have an `iInnerVal` member, and an `Inner` object does not have an `iOuterVal` member. Provided that `Inner` is a `public` member of class `Outer`, creating an object of class `Inner` is the same as any other class except that the scope resolution operator `Outer::` is used.

The classes have no special access to each other's members; in other words, the classes do not have a `friend` relationship. `Outer` cannot see `iInnerPrivateVal` and `Inner` cannot see `iOuterPrivateVal`.

A nested class is typically a `private` utility class that has semantics only relevant to the outer class. The outer class would create a member object of the inner class and use it directly.

Pointer to members

In the same way that C++ declares regular pointers, for example `int* p` is a pointer to `int`, it is possible to declare a pointer to a member:

```
class Foo
  {
public:
  int iVal;
  int Bar(int);
  };

// In use
int* p;              // A regular pointer to an int
int Foo::* pm;       // A pointer to an int member in Foo
```

```
pm = &Foo::iVal;       // Assigns the "address" of the member
Foo foo;               // An example of dereferencing

int i = foo.*pm;        // Copy the value of foo.iVal using the offset
                        // value in *pm
```

In this example, `Foo::*` is a pointer in the same way that `int*` is a pointer, but it also contains the scope resolution operator `::` and class name `Foo`, indicating this is a pointer to an `int` inside the *class scope* of `Foo`. To assign a value to the pointer, the address-of operator `&` is used to get the "address" of `Foo::iVal`.

Pointers to members don't actually contain the physical addresses of object members (`&foo.iVal`), but they contain enough information to locate the member for any given instance of the class. The method of doing this is implementation-dependent, but a pointer to a member typically contains the offset of the member within the class. To access the member in a specific object, the object address is augmented by the offset contained in the member pointer. (See virtual tables in Section 3.3.)

For completeness, here is the syntax for a pointer to a member function:

```
int (Foo::*pmf)(int) = &Foo::Bar;
// compared to a regular pointer to function:
int (*pf)(int) = &SomeFunction;
```

See Section 1.4 for further information about regular function pointers.

The `this` pointer

Any non-`static` member function (see the next subsection for more on `static`) has a hidden pointer called the `this` pointer, also referred to as *self-reference*. The `this` pointer contains the address of the specific object instance being invoked. It is implicitly passed as the first parameter in all member method calls, allowing the method to act on member data for that instance.

A developer can also use the `this` pointer directly, for example to prevent self-assignment in an assignment operator and to return the object.

```
class Foo
  {
public:
  Foo(int n=0){iVal=n;}               // Constructor

  Foo& operator=(const Foo& obj)    // Assignment
    {
    if (this!=&obj)           // Check it's not assigning itself
      iVal=obj.iVal;          // copy the data
```

```
    return *this;           // dereference the pointer to return Foo&
                            // reference. This allows chained assignments.
    }
private:
  int iVal;
  };

// In use
Foo foo(24);               // Straight declaration
foo = foo;                 // nothing much happens!
Foo footoo = Foo() = foo;  // Chain 1: Foo() is a temp, all contain 24
Foo fooTuTu=footoo=foo=Foo(42);  // Chain 2: all contain 42
```

The example shows a useful feature of the `this` pointer (and references to it). Two multiple-assignment statements (chains 1 and 2) demonstrate how values are passed through objects, even temporary objects.

Constant and static members

A `const` member method does not modify or change the state of the class in any way. A `const` member method is allowed to invoke only other `const` members to prevent it breaking the `const`ness, whereas non-`const` methods can invoke both `const` and non-`const` functions.

A `static` member is a variable, constant or function for which there exists only one copy per class, as opposed to non-static members, which are unique for each object. In effect, it is a single "global" variable or function for all objects in that class. Static functions do not have a `this` pointer and a class definition contains only the declaration of static data members; the definition occurs outside the class.

```
// Don't try this in Symbian OS DLLs - writable static data is not
// recommended!
class Foo
  {
public:
  // Static Methods
  static Foo* Create()
    {
    if (iObjectCount<iMax)
      {
      ++iObjectCount;
      return new Foo();
      }
    else
      return NULL;
    }

  static int ObjectCount()
    {
    return iObjectCount;
    }
```

```
  // destructor
  ~Foo(){--iObjectCount;}

  // Accessors and Modifiers
  int Get() const {return iVal;}    // const member i.e. doesn't modify
  void Set(int n) {iVal=n;}         // non-const modifies

private:
  // Static Declarations
  static int iObjectCount;          // Global static variable
  static const int iMax;            // Global static const or read only

  // Object variables
  int iVal;

private:
  // private constructor i.e. no direct newing of Foo, only use the
  // factory method
  Foo(){iVal=0;}
  };

// Static Definitions
int Foo::iObjectCount = 0;    // Note the static keyword is not required
const int Foo::iMax = 10;

// In use
Foo* p = Foo::Create();                // Calls the static method
if (p)
  {
  int count = Foo::ObjectCount();  // count == 1
  p->Set(42);
  int n = p->Get();
  delete p;
  count = Foo::ObjectCount();        // count == 0
  }
```

The `Foo` class contains two `static` functions. These functions can be called without the need to instantiate any objects, and the class scope resolution operator `::` is used to call them. `Create()` is a *factory* function used to create new `Foo` objects and return a pointer to them. `Create()` increments the static variable `iObjectCount` to count the number of objects it has manufactured and it stops when the limit defined by the `static const iMax` is reached. The constructor has been made private to stop `Foo` objects being directly created and to force users of the class to call the factory function. See Chapter 6 for more about the two-phase-construction idiom.

The destructor `~Foo()` decrements `iObjectCount`.

The static data members `iObjectCount` and `iMax` are declared inside the class scope and defined outside it.

Writable static data is not recommended in Symbian OS DLLs, but there are techniques to get around this. See Chapter 10 for more information.

Exceptions

- Nested class access differs between different versions of the C++ standard.
- Always exercise care when using `static` data members, especially when dealing with portability issues. Initialization of global `statics` is implementation dependent.

Exam Essentials

- Describe `private`, `protected` and `public` access control for class members
- Declare and specify the syntax for pointers to class members
- Identify nested classes and their scope and lifespan
- Cite the scope access rules and lifetime of a nested class
- Understand the scope rules and syntax of `friend` functions
- Understand the semantics of member functions and data addressing
- Understand the purpose of the scope-resolution operator `::`
- Understand the role of the `this` pointer
- Specify the properties of `static` class members

References

[Dewhurst 2005 Items 15, 16 and 30]
[Gamma *et al.* 1995 Chapter 3]
[Meyers 2005 Chapter 1 (Items 2 and 3)]
[Stroustrup 2000 Sections 10.2.4 and 10.2.6, Sections 11.5 and 11.12, Sections 15.3 and 15.5, and Sections C.9, C.10.2 and C.11]

3

Class Design and Inheritance

Introduction

Chapter 2 described the properties of classes and objects, introducing C++ support for the object-oriented programming (OOP) paradigms of abstraction, encapsulation and implementation hiding. This chapter introduces C++ provision of inheritance and both dynamic and static polymorphism.

Inheritance is a large and complex subject area, where innocuous and seemingly obvious facts often turn out to have far-reaching ramifications. Operator overloading, for example, has arguably led to more programming errors and inefficient code than the problems it solves. This is no reflection on the language itself but, more often than not, is caused by enthusiasm overtaking attention to detail.

This chapter does not give a complete specification of OOP inheritance but delivers the salient facts that an ASD candidate is expected to know and understand.

3.1 Class Relationships

C++ classes serve two primary purposes: to support user-defined types and to provide reuse through inheritance and composition relationships. This section describes the various OOP relationships.

Critical Information

Inheritance, composition and aggregation

There are fundamentally two kinds of relationship between classes in the C++ support of OOP: "is-a" or *inheritance,* and "has-a" or *composition.* Composition can be further divided into strong and weak or *aggregation.*

Relation	OO Term
S60 is a Symbian OS UI platform	Inheritance (is-a)
UIQ is a kind of UI platform	Inheritance (a-kind-of)
W950i is a mobile phone and is a music player	Multiple inheritance (is-a)
An N91 is part of the Nseries range	Aggregation (weak has-a)
A Smartphone device has an operating system	Composition (strong has-a)

*(*N91 and W950i are both Smartphones, Nseries is a product line)*

The table above contains some examples of these relationships: S60 is specifically a Symbian UI platform, and along with UIQ is more broadly referred to as a kind of UI – as would be, say, a PC or a car dashboard. The Sony Ericsson W950i is both phone and a music player. All three of these relationships are **inheritance** relationships as they either extend or specialize (and in the last case also combine).

If the N91 was removed from the Nseries product line, both the product line and N91 would still exist, the N91 would continue to be a phone (or multimedia computer) and the Nseries would still be a collection of multimedia devices. This is weak composition or **aggregation**.

Whereas a Smartphone device has to have an operating system to be a Smartphone, the operating system is an integral part of the Smartphone but the Smartphone is not an operating system, in other words there is no inheritance. This is strong **composition**: the operating system requires device hardware to run on and the device requires an operating system.

In the examples above, it is possible to see additional relationships, for example a UI is part of Symbian OS and UIQ is a Symbian OS UI platform. Good OO analysis and design require the careful selection and use of these relationships.

Reuse advantages and disadvantages

Aggregation, composition and inheritance are primarily used to remove redundancies, in particular the repetition of the same code (often implemented differently) in different classes or programs. In the analysis of a problem, if a group of concepts share common properties and behavior they are good candidates for abstraction, in this case, the separation of the common functionality. The amount of commonality dictates the degree of code *reuse* and the purpose of the relationships, that is, whether a class should inherit or contain the common entity.

The advantages of C++ class reuse:

- Reduced development time
- Fewer defects – this is a self-stoking cycle as the more a class is used, the more likely errors are to be uncovered and fixed
- Enforced interface consistency – a known interface at design time provides positive constraints on the design
- Known behavior
- Rapid prototyping with live code
- Faster implementation modifications – only a single class needs to be modified.

There are also disadvantages:

- Increased coupling between classes – modify a common class interface and there is a cascade effect or compatibility break (see Chapter 16)
- Design tradeoff problems – a large class can introduce unnecessary code bloat to client classes requiring only a specific aspect of that class
- Complexity – too many small base classes can introduce complex or multiple inheritance relationships
- Difficulties in understanding inheritance relationships – dynamic behavior is not always clear.

Exceptions and Notes

In OOD, types and classes are seen as different concepts: a type has its roots in set theory and formal methods, whereas classes are seen as structure concepts. Some languages, such as Smalltalk, maintain this separation; in C++, types and classes are handled the same. This can lead to some confusion when talking about relationships in a more formal context.

Exam Essentials

- Understand the key benefits and purpose of inheritance
- Specify the differences between composition, aggregation and inheritance
- Cite the object-oriented relationship for inheritance

References

[Ambler 2001 Chapters 5–7]
[Gamma *et al.* 1995 Chapter 4]
[Meyers 2005 Chapter 6]
[Morris 2006 Chapter 5]
[Stroustrup 2000 Section 24.3]
[Sutter and Alexandrescu 2004 Item 34]

3.2 Inheritance

The terminology used here is *base* and *derived* class where the base class is the parent or superclass and the derived class is the child or sub-class. *Class user* refers to code, or another class, that has no inheritance or friend relationship with the class and has only public access.

Critical Information

Public inheritance

Public inheritance is a relationship between derived and base classes where the derived class "is-a" specialization of the more generic base class.

The following example typifies a public base-derived inheritance relationship, showing the use of constructors to ensure the correct initialization of values from a derived class. It introduces the `protected` keyword as an inheritance access scope and the `virtual` member function.

```
// A read-only base class
class Base
  {
public:
  Base():iVal(0){}                    // Default constructor
  Base (int n):iVal(n){};             // Overload for value

  int Get() const {return iVal;}      // Access method (read only)

protected:                            // Only accessible to derived classes
  virtual void Set(int n){iVal=n;}    // write method
  void Reset() {Set(0);}              // will call the correct Set()

private:
  int iVal;                           // private data member
  };

// A read-write class:
// Derived overrides the protected Set() method in Base extending the
// functionality of the base class by allowing write objects.
class Derived : public Base           // Inherits the public and protected
                                      // interface
```

```
  {
public:
  // Default constructor
  Derived(bool write = false): Base(42), AllowWrite(write)
    {
    }

    // Override
  void Set(int n)      // Overrides the protected Set() method of Base
    {
    if (AllowWrite)
      Base::Set(n);
    }

private:
  const bool AllowWrite;  // const state i.e. only one state defined at
                          // initialization
  };

// In use
Derived foo(true);       // allow write
foo.Set(24);             // use Derived::Set() to set a value
int n = foo.Get();       // use Base::Get() to get the value
```

The derived class publicly inherits the base-class `public` interface but has no special access permission to the `private` area of the base class. To support a finer degree of access control for inheritance, C++ provides the `protected` keyword. This allows publicly inheriting derived classes access to member functions and data declared inside the `protected` sections of the base class, while still hiding them from class users. Use of the `protected` keyword tells the developer this is where the base class is expected to be extended (see Section 3.3 for the use of substitution and `virtual` member functions).

Base-class default constructors are implicitly invoked in the order of inheritance, before their derived-class constructors, that is before the body of the derived constructor executes. Base-class destructors are called in the reverse order after the body of the derived-class destructor has finished executing. (See Section 3.3 for the use of virtual destructors.)

Virtual methods

Virtual methods are covered in more detail in Section 3.3, but for better understanding of the role of inheritance, here is an overview.

A virtual method is a method declared in a base class which, provided the prototypes are identical, allows any derived class to redefine it or, more formally, *override* it. The `virtual` keyword tells the compiler to ensure the correct version of the function is invoked for the relevant derived class.

The purpose of *virtual overriding* is to allow member functions to be *dynamically bound* to their defining object class. In the previous example, the `Reset()` function in the `Base` class will call the redefined `Derived::Set()` and not `Base::Set()` because `foo` is created as a `Derived` class. A non-virtual override of `Set()` would result in `Reset()` calling `Base::Set(0))` regardless of the derived implementation, as it would be *statically bound* to `Base`.

Virtual overrides allow base-class run-time behavior to be *modified* by a derived class without the base-class implementation *seeing* the derived class. In fact the derived class can be added at a later date, such as a plug-in to a framework. This is sometimes referred to, somewhat hyperbolically, as "code written today to call code written tomorrow".

Note the use of the scope resolution operator `::` where the derived `Set()` function calls `Base::Set(n)`. Without the `Base::` qualifier, `Derived::Set()` would recursively call itself.

What is not inherited (who needs friends?)

Certain methods and properties are not inherited in a base-derived relationship:

- The base class constructors and destructors
- The assignment operator, =
- Friends – a `friend` of a base class has no special access permissions to a derived class (see information on friends in Section 2.3).

Private inheritance

Public inheritance is intuitive. The derived class has access to the `public` and `protected` members of the base class; the class user automatically has access to the base public interface while the `protected` members stay hidden. Public inheritance is the C++ representation of an "is-a" relationship.

The purpose of private and protected inheritance may be a little less obvious.

```
class Derived : public Base;    // public inheritance - as above

class Derived : private Base;   // private inheritance

class Derived : protected Base; // protected inheritance
```

In a private inheritance relationship, everything in the `Base` class is imported into the private area of the `Derived` class and is hidden from the users of the `Derived` class. The `Base` class in effect becomes a private class in the `Derived` class; thus the relationship is no longer "is-a" (inheritance) but "has-a" (composition).

Private inheritance is an implementation feature; there is no design-time or *conceptual* relationship between a privately inherited base class and a derived class. It is more accurate to represent private inheritance as the derived class "is-implemented-in-terms-of" the base class.

```
// Base class
class Radio      // GSM or CDMA Radio, not FM or bluetooth
  {
public:
  // many interface functions ...
  // ...
  virtual TState OnModemRequest()          // Virtual Method

  void ModemReq()
    {
    TState state = OnModemRequest();       // Override call
    switch (state)  // ... etc
    }
  };

// Derived class
class CellPhone : private Radio
  {
public:
  void MakeCall();

  // ... many interface functions ...
private:
  TState OnModemRequest();     // note the virtual keyword is not required
  };
```

In the example, the `Radio` class is a generic utility class, its methods are public and there is a `virtual` event method `OnModemRequest()` which requires some action on the part of a derived class.

A cell phone "has-a" radio, so at design time it makes perfect sense to use composition – the `CellPhone` class would have a `Radio` member object. But the `CellPhone` class requires notification of a `ModemReq` to tell the `Radio` its state, delivered via the virtual `OnModemRequest()` method; to override a method, the class has to be inherited.

If the `CellPhone` class publicly inherits `Radio`, its interface would be made available to users of the `CellPhone` class. This would make very little sense, as the `CellPhone` is not a `Radio` type (not "is-a") and it is no longer a "has-a" relationship, either. The `CellPhone` class "is-implemented-in-terms-of" the `Radio` class.

The advantage of privately inheriting is that it hides the `Radio` interface while maintaining the `CellPhone` logical interface; in other words, it maintains encapsulation while providing the implementation functionality that composition cannot support.

Protected inheritance

Protected inheritance, like private inheritance, is a property of implementation and hides the public interface of the base class. Protected inheritance applies `protected` access in the derived class to the `public` and `protected` members of the base class. This allows any classes (*grandchildren*) that further derive from the derived class (*child*) to have access to the base (*parent*) methods, unlike private inheritance, which stops any further access to the base class in the inheritance hierarchy.

There is rarely a valid reason for using protected inheritance.

Note: neither private nor protected inheritance prevents virtual functions from being overridden.

Inheritance summary

- Private inheritance prevents both public and any further inherited access to base members.
- Protected inheritance prevents public access but allows further derived classes access.
- Friends of base classes are not friends of derived classes.
- Neither private nor protected inheritance prevents virtual functions from being overridden.

Exam Essentials

- Be able to define public inheritance
- Understand the scope resolution operator syntax for accessing the base-derived class hierarchy
- Given a base-derived hierarchy, specify the access rules (including those of friend classes)
- Describe the scope access rules and purpose of public, protected and private inheritance
- Specify the implicit invocation order of constructors and destructors in a base-derived hierarchy

References

[Meyers 2005 Chapter 6 (Items 38 and 39)]
[Stroustrup 2000 Section 12.2 and Section 15.3]
[Sutter 1999 Items 22 and 24]
[Sutter and Alexandrescu 2004 Items 34–7]

3.3 Dynamic Polymorphism – Virtual Methods

This section looks at C++ support for *polymorphism,* specifically *dynamic* or *run-time* polymorphism, and related implementation issues. C++ also supports *parameterized* or *compile-time/static* polymorphism, which is covered in Section 3.4.

Unless otherwise stated, all inheritance relationships are public.

Critical Information

Polymorphism defined

The definition of polymorphism, along with its ubiquitous shape example, is possibly one of the most well-trodden in C++. A simplification would be: objects of different (but related) types can respond to the same method without the caller of the method needing to know the type of object that defines the method.

Thus a square, a circle and a triangle are all types of shape. A developer wishing to write a program to render them to a screen makes the shape type polymorphic by declaring a shape base class with a virtual draw method and providing all the derived shape classes (squares, etc.) with a redefined draw method which is called from within a single rendering routine.

Substitution

The key property of any programming language that supports dynamic or run-time polymorphism is that it should be possible to replace a base-class object with a derived-class object without affecting the behavior of the code which calls the interface. This is known as *substitution*, specifically referred to as the "Liskov substitution principle" after its inventor.

To support substitution, and good class design, a class is *specialized* by extending its interface and should not put any additional constraints, or stronger *preconditions*, on the use of the base-class interface functions. For example, if in the earlier example of read-only and read/write classes (see "Public Inheritance" in Section 3.2) `Base` was writable and `Derived` was read-only, the behavior of the calling code would be affected, as the writable `Set` method would no longer be available in the derived (read-only) class.

On the other hand, *post-conditions* may be strengthened. For example, if only modified values were returned from the `get()` method, the behavior of the calling program would not be affected as it would only act on a narrower band, or specialization, of values.

Dynamic binding and static strong typing

C++ is said to be a "statically strongly typed" language, that is, declarations and expressions are checked for correct usage at compile time. But dynamic polymorphism requires dynamic binding, where the exact object type is not known at compile time.

The solution to this dilemma is that for an object to be polymorphic, it has to be handled via either a pointer or a reference.

```
class Base
  {
public:
  virtual void Foo();
  };

// Derived class ...
class Derived : public Base
  {
public:
  void Foo();
  };

//
void SomeNonMember (Base* p)      // no binding to Derived
  {
  p->Foo();                       // calls the correct virtual override
  }

// calling
Base* p = new Derived();          // p is bound to Derived
SomeNonMember(p);
```

Section 3.2 stated that virtual methods ensure the correct version of a derived method is called within the base class, thus allowing a derived class to extend the base class with any modification.

In the same way, in the above example the `SomeNonMember()` function needs to deal only with `Base` pointers and the virtual override mechanism ensures the correct `Foo()` method is called – `Derived::Foo()`. If at a later stage a `DerivedTwo` class was also created in the same manner, `DerivedTwo::Foo()` would be invoked without any modification to `SomeNonMember()`.

It is possible for `delete` to be called on the `Base` pointer; this causes only the `Base` destructor to be called, leaking or orphaning any memory owned by the `Derived` class. Thus, any class that has at least one `virtual` method should declare its destructor `virtual`, ensuring that the derived destructor is called (the base destructor will be called after the derived destructor as normal).

Overrides and overloads

These terms can be a source of confusion so they have a short section of their own. Virtual overriding is a property of C++ support for dynamic binding and polymorphism, and operates *vertically* in a class hierarchy: `D::Foo()` overrides `B::Foo()`.

Overloading operates *horizontally* in a single class scope. Functions that share the same name but have different parameter lists are said to be overloaded. `A::Set(int)` and `A::Set(char)` are overloads of a `Set()` function.

Overriding and overloading are completely different and share no relationship with each other.

Overloading

Overloading allows a developer to provide more than one version of a function or operator within the same scope. Operator overloading allows user-defined types to behave in the same manner as integral types. Arguably, overloading provides a loose ad-hoc form of polymorphism via the same method having different types, but it has no claim to true OO polymorphism and is seen as a notational convenience.

```
class Foo
  {
public:
  // Constructor
  Foo(int n = 0):iVal(n){}

  // Overloaded functions
  int Get() const {return iVal;}          // Differentiate on parameter
  void Get(int* val) const {*val = iVal;} // lists, not on return type

  // Overloaded operators
  Foo& operator+=(int n)          // Add an int
    {
    iVal+=n;
    return *this;                 // return (see this pointer section)
    }

  Foo& operator+=(const Foo& foo) // Add a Foo
    {
    iVal+=foo.iVal;
    return *this;
    }

  // conversion (or cast) operator:
  operator int&()                 // converts to an int(ref)
    {
    return iVal;
    }

private:
  int iVal;
  };

// a binary overload outside class scope (value return)
Foo operator+(const Foo& a, const Foo& b)
  {
  Foo res = a;                // default initialization
  return res+=b;              // Uses operator+=(const Foo&) overload
  }
```

```
// In use ...
Foo foo(42);
foo+=100;                           // foo.iVal == 142
Foo bar(1000);
foo+=1000;                          // foo.iVal == 1142

int n = foo.Get();                  // n == 1142
int m = 0;
bar.Get(&m);                        // m == 1000
Foo foobar;
foobar = foo+bar;                   // foobar.iVal == 2142

// Conversion using the operator int&()
int val = static_cast<int>(foo);        // explicit compile-time
val = foo;                              // automatically call
```

Note that the conversion operator `int&()` may used in conjunction with the `static_cast` operator or called automatically as part of the integral `int` assignment operator.

The examples also include an example of the binary operator + which is defined outside the class but implemented using the member operator +=. The binary operator could be defined within the class but with one fewer argument:

```
Foo Foo::operator+(const Foo& foo)
  {
  Foo res = *this;              // default initialization
  return res+=foo;              // Uses operator+=(const Foo&) overload
  }
```

The rules of overloading can become a little tricky. Here are the more common ones:

- Unary operator overload members take no arguments
- Binary operator overload members take one argument
- Non-member unary operators take one argument and non-member binary operators two arguments
- Ternary member operators are not supported
- The parameter list determines which overload is called and the return type is ignored
- Precedence and associativity of operators cannot be changed
- All overloaded operators can be inherited except the assignment operator
- It is not possible to overload operators using only built-in or integral types
- It is not possible to create completely new operators (for example operator#)

- It is not possible to change the arity (number of arguments) of any operator, for example a unary operator must remain unary.

Operators that may be overloaded:

<table>
<tr><td>Unary operators</td><td>+ - * & ~ ! ++ -- -> ->*</td></tr>
<tr><td>Binary operators</td><td>+ - * / % ^ & | << >> += -= *= /= %= ^ = &= |= <<= >>= < <= > >= == != && || , [] () new new[] delete delete[]</td></tr>
</table>

Operators that may not be overloaded: `. .* ?: ::`

Overload function-parameter-matching strategies in order of precedence:

Exact Match	Including simple conversions (e.g. array to pointer, non-`const` to const)
Match using promotion	Integral type-safe (smaller to larger) promotions (`char` to `int`, `short` to `int` etc.)
Match using standard conversions	Includes `derived*` to `base*`, `int` to `unsigned`, `p*` to `void*`
Match using user-defined conversions	Another already user-defined operator e.g. `operator int& Foo` to `int`. See above example.
Match using ellipses	Matching using `...` is a last-ditch attempt i.e. any number of arguments

Any ambiguous match at the same level will be marked as an error, for example `Bar(T& ref)` and `Bar(T val)`. Calling with a value of type T is seen as an exact match for `Bar(T val)`, but because C++ treats references identically to values in this context, it is also an exact match with `Bar(T& ref)` and the compiler will flag it as an error.

Virtual table

To support dynamic binding, and thus polymorphism, C++ uses a virtual table (*vtable*) to resolve the correct derived virtual method from a base-class pointer. The vtable is created in classes that declare at least one virtual function.

The vtable of a base-class object contains an array of pointers (offsets or addresses) to each of the derived-class overridden function implementations. Thus a pointer to a base-class object calls the correct derived method by dereferencing the pointer to the method.

Only the base-class object contains the vtable, with each derived-class object containing a pointer (*vptr*) to the table. When a method is called in a derived class, the vptr is used to get the correct method address from the vtable. This is an overhead compared to a non-virtual member method, which is statically linked by the linker and may be called directly.

Multiple inheritance and abstract base classes

The use of multiple inheritance is seen as controversial in general C++ and is definitely not encouraged in Symbian OS C++ except under specific circumstances. When a derived class derives from two bases, member name ambiguities can arise; in the case of diamond inheritance, where the base classes both derive from a common base, there is a complete duplication of data.

```
class B
  {
protected:
  int iVal
  };

class D1 : public B{};
class D2 : public B{};
class MI : public D1, public D2
{...} // which iVal ?
```

The class now has two versions of `B::iVal`.

Symbian OS C++ uses multiple inheritance purely as an interface inheritance mechanism. These interface classes are called *mixins* (see Chapter 4) and use the C++ language's support of *abstract base classes*.

An abstract base class is a class that cannot be instantiated, that is no object of that class may be directly created. Its primary purpose is to allow an interface to be declared without providing any implementation. Although it is possible for abstract classes to contain data members, this is not recommended.

For a class to be abstract it must contain at least one *pure virtual* member method:

```
class Interface     // Abstract Base Class
  {
public:
  virtual void Foo()=0;          // Pure virtual method declaration
  // ...
  };
```

```
class Impl : public Interface    // Derived
  {
public:
  void Foo(){...}                // Definition
  };
```

All pure virtual methods must be defined in the derived class, otherwise the derived class itself is abstract and cannot be instantiated.

In the case of multiple inheritance, it is valid to publicly inherit an abstract base class (interface), and privately inherit an implementation class – privately because it is good design to hide the implementation class interface from the user (see information on private inheritance in Section 3.2). It is also valid to inherit a number of abstract base class interfaces.

Interface and implementation inheritance

From the above sections it is possible to see the difference between inheriting an *interface*, or abstract base class, and inheriting an *implementation*. To summarize:

- Public inheritance means the interface is always inherited
- Pure virtual methods indicate the class is abstract and is an interface which must implemented in a derived class
- Ordinary virtual methods are interface functions that maybe overridden, but the base class provides a default implementation
- Non-virtual methods are the interface and mandatory implementation of the base class
- Private inheritance means the interface is not exposed to class users but is visible to the derived class; this is known as implementation inheritance.

Exceptions

Mixins may contain actual implementation code for common default behavior.

Exam Essentials

- Specify the mechanisms of OO reuse available in C++
- Be to able to state C++ support for polymorphism

- Understand the purpose of and difference between overriding and overloading
- Understand the use of overriding to modify behavior in base-derived class inheritance
- Understand the rules and pattern-matching criteria for correct overloaded-function invocation
- Identify the typical uses and behavior for operator overloading
- Describe the purpose of the virtual table, citing constraints and overheads
- Specify the use of virtual functions and their implementation tradeoffs
- Understand how an abstract base class is implemented in C++
- State the differences between interface and implementation inheritance
- Understand and recognize the problems associated with multiple inheritance
- Cite the implementation requirements to support the `static_cast` operator in user-defined classes

References

[Ambler 2001 Chapter 7]
[Dewhurst 2005 Items 2, 21 and 33]
[Meyers 2005 Items 34, 40]
[Morris 2006 Chapter 5]
[Stroustrup 2000 Chapter 11, Section 12.2.6 and Section 15.2]
[Sutter 1999 Items 21 and 24]
[Sutter and Alexandrescu 2004 Items 26–31]

3.4 Static Polymorphism and Templates

The previous section focused on the use of virtual overrides for dynamic polymorphism. This section looks at static or compile-time polymorphism using C++ parameterized types called templates. Although standard C++ templates are not widely used in Symbian OS development, an Accredited Symbian Developer is expected to understand the basic terminology and semantics of C++ templates. Symbian OS thin templates are covered in the final part of this section.

Critical Information

Template functions

Templates can be tricky things to talk about, primarily because they fall between C++ programming and programming the compiler itself. A template is a function or a class that is decoupled from the specific type, or types, it acts on. The template is provided with a parameter list containing type names that allow the compiler to create type-specific instances of the function or class.

```
// template function for a simple addition function
template<typename T>            // template parameter list
T Add(T a, T b)                 // function template name
  {
  return a+b;
  }

// int example
int a = 2;
int b = 2;
int res = Add(a,b);             // i.e. int Add(int, int)

// float example
float c = 2.2;
float d = 2.2;
float res2 = Add(c,d);          // i.e. float Add(float, float)
```

The `template` keyword is followed by the template parameter list, in which `<typename T>` tells the compiler to substitute `T` with the type of the argument being passed when it is generating code using the template.

Thus when the compiler encounters the `int` in the first example the code for `int Add(int a, int b)` is generated, and for the second example `float Add(float a, float b)`. Note the template parameter list maybe also be declared using "`class`" instead of "`typename`" that is `template<class T>`. These names are interchangeable when declaring a template parameter.

The reuse of the same generic code for different types is polymorphism. As the code is generated at compile time, and statically bound to the type passed in the template parameter list, it is known as static or compile-time polymorphism. The generated code allows calls to the function to be type-safe, a great advantage over using preprocessor macros, which have no type checking. It is also generally easier to debug templates than macros.

Template classes

As well as function templates, C++ supports class templates.

```
// template class.
template<typename T>              // The parameter list
class Foo                         // The template name id
  {
public:
  // constructor
  Foo(){};                       // default
  Foo(T val):iVal(val){}

  // Modifiers & Accessors
  void Set(T val){iVal=val;}
  T Get()const {return iVal;}

private:
  T iVal;
  };

// in use
Foo<int> foo(0);                  // Builds an int iVal class
Foo<float> boo(0.0);              // Builds a float iVal class
```

The semantics for template classes are the same as for template functions. Here two classes are generated from the template definition, one for dealing with an `int` and one for dealing with a `float`. Template classes may inherit other template classes:

```
template <typename T, typename C>
class Bar : public Foo<T>         // this is using a template
  {
public:
  // ...

private:
  C iDerivedVal;
  };

// in use
Bar<int, float> bar;
bar.Set(42);
```

The first parameter (`T`) is passed to the base class `Foo` and the second (`C`) is used by the derived class `Bar`.

Template classes are the same as any other class in many respects:

- They can have friends – a friend will be a friend of every class created from the template
- They have statics – each class created from the template has its own unique static, that is `class<int>` and `class<float>` will have separate static globals
- They can have nested classes

- They can inherit from non-templates
- They can have overrides and overloads.

Template specialization

In the template function example of `Add()`, above, the heart of the template is the addition expression `a+b`, This means the template makes sense for any type where `a+b` is semantically correct, including user-defined classes that override the addition operator `+`. But there are exceptions where adding is a valid operation but there is no `operator+` defined, or a function is required: for example, the `char*` type uses the function `strcat()` to add two strings together.

```
// Generic
template<typename T>                        // template parameter list
T Add(T a, T b)                             // function template name
  {
  return a+b;
  }

// Specialization
// The empty parameter list tells the compiler it is dealing with
// a specialization:
template<>
char* Add<char*>(char* a,char* b)           // Template argument list
  {
  return strcat(a,b);
  }

// In use
char* res = Add<char*>("foo","bar");        // note different syntax
int n = Add(2,2);                           // calling the int version
```

The empty `template<>` parameter list indicates to the compiler to expect a template specialization. An important note about terminology and syntax: the angle brackets in the statement `char* Add<char*>(char* a, char* b)` indicate an argument list and not a parameter list as in the declaration. They tell the compiler to create this version of the template when the `<char*>` type is encountered. The argument list must also be contained in the function call itself (`Add<char*>`) to ensure the compiler creates the correct version.

Non-type parameters

It is also possible to pass non-type (that is, value) parameters to templates. These are typically const values.

```
template<typename T, int size>
```

```
class Foo
  {
  T iElements[size];        // an array of Ts
  };

// In use
Foo<int,10> foo;              // Creates an array of 10 integers
```

Introducing Symbian OS thin templates

Symbian OS provides its own variation of templates. This is partly because compilers at the time of EPOC's inception lacked sufficient template support. More importantly, templates may lead to code duplication if not used correctly.

The Symbian thin template idiom works by implementing the essential functions (for example `Pop()` and `Push()` for a stack template) in a protected base class and using `void*` (`TAny*`) pointers to provide *type-agnostic* behavior. The base class is privately inherited (see Section 3.2) with the derived class using a regular C++ template to specify the type.

```
class StackBase                   // Implementation class
  {
protected:                        // template interface *hidden*
  void Push(void* item);
  void* Pop();

private:
  void* iStack[100];            // or some other internal representation
  };

template<typename T>
class Stack : private StackBase
  {
public:
  void Push(T* item) {StackBase::Push((void*)item);}  // must be inline
  T* Pop() {(T*)StackBase::Pop();}                     // must be inline
  };
```

The `Stack` class implements its functions inline to reduce the class overheads, only writing the code where it is needed. In effect the `Stack` template is a type-safe wrapper for the underlying type-agnostic `StackBase` container class.

Exceptions and Notes

Friend template classes may be specialized for specific classes; for example inside a class template declaration, `friend class Bar<T>;` binds that class to be only a friend of classes of type `T`.

Exam Essentials

- Specify the syntax for a simple function template specialization
- Be able to cite the advantages of function templates (for example over macros)
- Understand the inheritance rules and syntax supported by class templates
- Understand the syntax and semantics of a template type/class declaration
- Recognize the prototype declaration and pattern-matching properties of a template declaration and its use
- Understand the purpose and implementation differences of the Symbian OS thin template and mainstream C++ templates

References

[Alexandrescu 2001 Forewords (Meyers, Vlissides), Chapter 1]
[Dewhurst 2005 Items 45–9]
[Meyers 2005 Item 41]
[Stichbury 2004 Chapter 19]
[Stroustrup 2000 Section 2.7 and Chapter 13]
[Sutter 1999 Item 23 Part 2]
[Sutter and Alexandrescu 2004 Items 64–7]

4

Symbian OS Types and Declarations

Introduction

Symbian OS defines a set of fundamental types which, for compiler independence, are used instead of the native built-in C++ types for compiler independence.

Symbian OS also defines several class types, each of which has different characteristics. These types are used to describe the main properties and behavior of objects of each class, such as where they may be created (on the heap, on the stack or on either) and how they should be cleaned up.

Because each of the class types has a well-defined set of rules, the creation, use and destruction of objects of each class type is more straightforward. To enable each of the types to be easily identified, Symbian OS uses a simple naming convention which prefixes the class name with a letter (T, C, R or M). The only classes which don't adopt one of these prefixes are those which possess only static member functions.

This chapter begins by discussing the fundamental Symbian OS types, and then describes the characteristics of each of the different class types. It concludes by summarizing which factors are important when writing a class and deciding which type to choose, and why it is important to follow the naming convention and rules for each type of class.

4.1 The Fundamental Symbian OS Types

Symbian OS provides a set of `typedefs` for the built-in C++ native types. These are guaranteed to be compiler-independent and should always be used instead of the native types.

Critical Information

The fundamental Symbian OS types are as follows.

- TIntX and TUintX (where X = 8, 16 and 32) are used for 8-, 16- and 32-bit signed and unsigned integers respectively. Unless there is a good reason to do so, such as for size optimization or compatibility, the non-specific TInt or TUint types should be used, corresponding to signed and unsigned 32-bit integers, respectively.
- TInt64: Releases of Symbian OS prior to v8.0 had no built-in support for 64-bit arithmetic on hardware builds, so the TInt64 class implemented a 64-bit integer as two 32-bit values. On Symbian OS v8.0 and later, TInt64 and TUInt64 are typedef'd to long long and use the available native 64-bit support.
- TReal32 and TReal64 (and TReal, which equates to TReal64) should be used for single- and double-precision floating-point numbers, equivalent to float and double respectively. Operations on these types are likely to be slower than those on integers, so they should be avoided unless they are absolutely necessary.
- TAny* should be used in preference to void*, effectively replacing it with a typedef'd "pointer to anything". TAny is thus equivalent to void but, in the context where void means "nothing", it is not necessary to replace the native void type. Thus a function taking a void* pointer (to anything) and returning void (nothing) will on Symbian OS have a signature as follows:

  ```
  void TypicalFunction(TAny* aPointerParameter);
  ```

  and not

  ```
  TAny TypicalFunction(TAny* aPointerParameter);
  ```

- TBool should be used for Boolean types. For historical reasons TBool is equivalent to int and the Symbian OS typedef'd values of ETrue (= 1) and EFalse (= 0) should be used. However, since C++ will interpret any non-zero value as true, direct comparisons with ETrue should not be made.

Exceptions

Always use void when a function or method has no return type, instead of TAny. This is an exception to the rule of replacing a native type with a Symbian OS typedef, because void is effectively compiler-independent when referring to "nothing".

Exam Essentials

- Know how the fundamental Symbian OS types relate to native built-in C++ types

Symbian type	Native type
TAny*	void*
TBool	bool
TInt	int
TUint	uint
TReal	float
TInt64	long long

- Understand that the fundamental types should always be used in preference to the native built-in C++ types (bool, int, float, etc.) because they are compiler-independent

4.2 T Classes

Critical Information

T classes behave much like the C++ built-in types, hence they are prefixed with the same letter as the typedefs described above (the "T" is for "Type").

Just like the built-in types, T classes do **not** have a destructor. In consequence, T classes must not contain any member data which itself has a destructor. T classes contain all their data internally and have no pointers, references or handles to data, unless it is owned by another object. Thus a T class will contain member data which is either:

- Built-in types and objects of other T classes
- Pointers and references with a "uses-a" relationship rather than a "has-a" relationship, which would imply ownership. A good example of this is the TPtrC descriptor class, discussed in Chapter 7.

Without a destructor, an object of a T class can be created on the stack and will be cleaned up correctly when the scope of that function exits, either through a normal return or a leave. The important thing to remember is that Symbian OS leaves (see Chapter 5) do not emulate the standard C++ throw semantics and the destructors of stack-based objects are **not** called when a leave occurs.

If a call to a destructor were necessary for the object to be safely cleaned up, the object could only be created on the stack in the scope of code which is guaranteed not to leave. This limitation would be too restrictive and risk causing a memory leak if overlooked. Thus the rule for T classes is that they must not need a destructor, so they can always be created and used on the stack and remain leave-safe.

An object of a T class can also be created on the heap. If it is a local variable, such an object should be pushed onto the cleanup stack prior to calling code with the potential to leave. In the event of a leave, the memory for the T object is deallocated by the cleanup but no destructor call is made.

T classes are also often defined without default constructors; indeed, if a T class consists only of built-in types, a constructor would prevent member initialization as follows:

```
TMyPODClass local = {2000, 2001, 2003};
```

Some T classes have fairly complex APIs, such as the lexical analysis class `TLex` and the descriptor base classes `TDesC` and `TDes`. In other cases, a T class is simply a C-style `struct` consisting only of public data.

The T prefix is used for enumerations too, since these are simple types. For example:

```
enum TMonth {EJanuary = 1, EFebruary = 2, ..., EDecember = 12};
```

Exceptions

In older Symbian OS code, C++ `structs` are sometimes prefixed with S instead of T. More recent Symbian OS code tends to define these as T classes.

Exam Essentials

- Know the purpose of a T class, what types of member data it may and may not own, and that it must never have a destructor
- Know what types of function a T class may have
- Understand that a T class may be created on the heap or stack
- Understand that a T class may be used as an alternative to the traditional C/C++ `struct`
- Know that the T prefix is also used to define an `enum`.

4.3 C Classes

Critical Information

C classes are suitable for objects that ought to be allocated on the heap. Their purpose is to contain and own pointers to large objects. C-class objects are frequently large objects in their own right, and are thus unsuitable for creation on the stack.

C classes must ultimately derive from class `CBase` (defined in `e32base.h`). This class has two characteristics which are inherited by its subtypes and thus guaranteed for every C class: safe construction and destruction, and zero initialization.

Safe construction and destruction

`CBase` has a virtual destructor, so a `CBase`-derived object may be destroyed properly by deletion through a base-class pointer. The virtual destructor in `CBase` ensures that C++ calls the destructors of the derived class(es) in the correct order (starting from the most derived class and calling up the inheritance hierarchy). Unlike T classes, it's fine for a C class to have a destructor defined, and they usually do, so C classes may have member variables which need cleaning up in a destructor, such as heap-based data or resource handles.

This kind of member data often needs to call leaving code (such as allocation or initialization functions) when it is instantiated so, in effect, a C class will need to call leaving code when it is created. However, constructor code should never be able to leave, because this can cause memory leaks.

To avoid this, Symbian OS uses *two-phase construction* for C classes. Two-phase construction is a pattern characterized by making all constructors private or protected, and instead providing a static factory function, usually called `NewL()` or `NewLC()`. Any construction code which may leave is called within the factory method, which guarantees leave-safety. Two-phase construction is described in more detail in Chapter 6.

Besides a virtual destructor, `CBase` also declares a private copy constructor and assignment operator. Their declaration prevents calling code from accidentally performing invalid copy operations on C classes; if they are not declared, the compiler will generate implicit versions which simply perform shallow copies of any member data. If a C class does need a copy constructor or assignment operator, it must, therefore, be declared and defined explicitly.

Zero initialization

The second key characteristic of `CBase`, and hence its derived classes, is that it overloads operator `new` to zero-initialize an object when it is first allocated on the heap. This means that all member data in a `CBase`-derived object will be zero-filled when it is first created, and this does not need to be done explicitly in the constructor. Zero initialization will not occur for stack objects.

For this reason, among others such as managing cleanup in the event of a leave, objects of a C class must **always** be allocated on the heap.

Exam Essentials

- Recognize that a C class always derives from `CBase`
- Know the purpose of a C class, and what types of data it may own
- Understand that a C class must always be instantiated on the heap
- Know that a C class uses two-phase construction and has its member data zero-filled when it is allocated on the heap
- Understand the destruction of C classes via the virtual destructor defined in `CBase`

4.4 R Classes

Critical Information

The "R" which prefixes an R class indicates that it owns an external resource handle, for example a handle to a server session. Within Symbian OS, the types of resource handles owned by R classes vary from ownership of a file server session (class `RFs`) to ownership of memory allocated on the heap (class `RBuf` or `RArray`). It is less common to write R classes than C, T or M classes, unless implementing a client–server framework (as described in Chapter 11), since R classes are usually used to store client-side handles to server sessions.

The resource handle will not be initialized in its constructor, since initialization may fail and a constructor cannot return an error or leave. Unlike for C classes, there is no equivalent `RBase` class which zero-fills the object on construction, so a typical R class will just have a simple constructor which sets the resource handle to zero, indicating that no resource is currently associated with the newly constructed object.

To initialize the R-class object with a usable resource, the class typically has a function such as `Open()`, `Create()` or `Initialize()` which can be called after construction to set up the associated resource and store its handle as a member variable of the R-class object.

An R class also has a corresponding `Close()` or `Reset()` method, which releases the resource and resets the handle value to indicate that no resource is associated with the object. Although in theory the cleanup function can be named anything, by convention it is almost always called `Close()`.

A common mistake when using R classes is to forget to call `Close()` or to assume that there is a destructor which cleans up the owned resource. This can lead to serious memory leaks, and is something to look for when performing a code review.

R classes are often small, and usually contain no other member data besides the resource handle. It is rare for an R class to have a destructor – it generally does not need one because cleanup is performed in the `Close()` method. This is because it makes sense for a client to release a resource handle as soon as it no longer needs it, rather than to wait for the resource to be cleaned up at a later time.

R classes may exist as class members, as automatic variables on the stack, or occasionally on the heap. In any case, the resource must be released in the event of a leave, typically by using the cleanup stack.

It is quite rare to define an R class, but it is necessary for client-side access using a handle to the server session.

Exam Essentials

- Know the purpose of an R class, to own a resource
- Understand that an R class can be instantiated on the heap or the stack
- Understand the separate construction and initialization of R classes
- Understand the separate cleanup and destruction of R classes, and the consequences of forgetting to call the `Close()` or `Reset()` method before destruction

4.5 M Classes

Critical Information

The ''M'' prefix stands for ''mixin'', which is a term originating from an early object-oriented programming system, where the mixin class is used for defining interface classes.

An M class is an abstract interface class which declares pure virtual functions. A concrete class deriving from such a class typically inherits from `CBase` (or a `CBase`-derived class) as its first base class and from one or more M-class ''mixin'' interfaces, and implements the interface functions.

The correct class-derivation order is always to put the `CBase`-derived class first, to emphasize the primary inheritance tree. It also enables C-class objects of derived classes to be placed on the cleanup stack using the correct `CleanupStack::PushL()` overload.

```
class CCat : public CBase, public MDomesticAnimal{...};
```

and not

```
class CCat : public MDomesticAnimal, public CBase{...};
```

On Symbian OS, M classes are often used to define callback interfaces or observer classes.

The use of multiple interface inheritance, as shown in the previous examples, is the only form of multiple inheritance encouraged on Symbian OS. Other forms of multiple inheritance can introduce significant levels of complexity, and the `CBase` class was not designed with this in mind.

An M class must have no member data. Since an M class is never instantiated and has no member data, there is no need for an M class to have a constructor. In general, careful consideration must be given as to whether an M class should have a destructor (virtual or otherwise). A destructor places a restriction on how the mixin is inherited, forcing it to be implemented only by a `CBase`-derived class. This is because a destructor means that `delete` will be called, which in turn demands that the object cannot reside on the stack, and must always be heap-based. This implies that an implementing class must derive from `CBase`, since T classes never possess destructors and R classes do so only rarely.

In general, a mixin interface class should not be concerned with the implementation details of ownership. If it is likely that client code will own an object through a pointer to the M-class interface, as described above, it is necessary to provide a means for the owner to clean up the object when it is no longer needed. However, this need not be limited to cleaning up through a destructor. It may be preferable to provide a pure virtual `Release()` method so the owner can just say "I'm done". It's then up to the implementing code to decide what this means (for a C class, the function can just call `"delete this"`). This is a more flexible interface because the implementing class can be stack- or heap-based, and can perform reference counting, special cleanup or other tasks. It isn't essential to call the cleanup method `Release()` or `Close()`, but it can help calling code to do so. First of all, it's recognizable and its function is easy to guess. More importantly, it enables the code using the class to make use of the `CleanupReleasePushL()` or `CleanupClosePushL()` functions described in Chapter 5.

Exceptions

An M class should usually have only pure virtual functions. However, there may be cases where non-pure virtual functions may be appropriate. A good example of this occurs when all the implementation classes of that interface have common default behavior. Adding a shared implementation to the interface reduces code duplication and the related bloat and maintenance headaches.

Exam Essentials

- Know the purpose of an M class, to define an interface

- Understand the use of M classes for multiple inheritance, and the order in which to derive an implementation class from C and M classes
- Know that an M class should never contain member data and does not have constructors
- Know what types of function an M class may include, and the circumstances where it is appropriate to define their implementation
- Understand that an M class cannot be instantiated

4.6 Static Classes

Critical Information

There are some Symbian OS classes which take no prefix letter and simply provide utility code through a set of static member functions, for example `User`, `Math` and `Mem`. The classes themselves cannot be instantiated; their functions must instead be called using the scope-resolution operator.

```
User::After(1000); // Suspends the current thread for 1000 microseconds
Mem::FillZ(&targetData, 12); // Zero-fills 12-byte block starting
                             // from &targetData
```

A static class is sometimes implemented to act as a factory class.

Exam Essentials

- Know that static classes do not have a prefix letter
- Understand that static classes cannot be instantiated because they contain only static functions

4.7 Factors to Consider when Creating a Symbian OS Class

Critical Information

It is important to consider the type of member data a class will contain, if any, when deciding what Symbian OS class type it is.

First, there are some cases where a class will not contain any member data. This may be because it is an interface class for inheritance only. This type of class will usually be defined as either an M class or, occasionally, a C class which has pure virtual or virtual functions only.

Another type of class which has no member data is a factory or utility class containing only static functions, in which case it is assigned no prefix at all.

If the member data has no destructor and needs no special cleanup (that is, if it contains only native types, other T classes, or pointers and references to objects owned elsewhere), it will typically be defined as a T class. However, this isn't appropriate if the size of data the class contains will be large (say, over 512 bytes). T classes can be created on the stack, and it's good to avoid large stack-based objects on Symbian OS, since space is limited. In cases like this, a T class can be defined, but it should be made clear that it should always be used on the heap. However, it is preferable to mandate that it is always created on the heap by making it a C class.

C-class objects are always created on the heap, and a C class should be defined if the class will own data which needs to be cleaned up, such as heap-based buffers, other C-class objects or R-class objects.

Exam Essentials

- Know the important factors to consider when creating a new class, and how this determines the choice of Symbian OS class type

4.8 Why Is the Symbian OS Naming Convention Important?

Critical Information

The class types simplify matters. The required behavior of a class can be considered and matched to the definitions of the basic Symbian OS types. Once this is decided, the role of the class can be considered. If the naming convention is followed, a user of an unfamiliar class can be confident in how to instantiate an object, use it and then destroy it in a leave-safe way.

Exceptions

Some Symbian OS classes do not conform to the standard Symbian OS class-type naming conventions. A good example is the `HBufC` class, described in Chapter 7.

Exam Essentials

- Understand that the use of a class prefix makes it clear to anyone wishing to use a class how it should be instantiated, used and destroyed safely.

- Recognize that the naming convention forces a class designer to think about the factors described in Section 4.7 and, having decided on the fundamental behavior, can concentrate on the role of the class, knowing that leave-safe construction, destruction and ownership are already handled.

References

[Babin 2005 Chapter 4]
[Harrison 2004 Chapter 1]
[Stichbury 2004 Chapter 1]

5

Leaves and the Cleanup Stack

Introduction

This chapter covers one of the most fundamental features of Symbian OS: leaves (Symbian OS terminology for exceptions) and the use of the cleanup stack to manage memory and other resources in the event of a leave. Symbian OS was designed to perform well on devices with limited memory and uses the cleanup stack to ensure that memory is not leaked, even under error conditions.

The chapter concludes with a short section on how to test for memory leaks in code using a set of debug macros supplied by Symbian OS.

5.1 Leaves: Lightweight Exceptions for Symbian OS

Critical Information

Symbian OS was first designed at a time when exceptions were part of the C++ standard, though not yet implemented in any compiler. Later, exception-handling support was found to add substantially to the size of compiled code and to run-time RAM overheads, regardless of whether or not exceptions were actually thrown. For these reasons, standard C++ exception handling was not considered suitable to add to Symbian OS, with its emphasis on a compact operating system and client code. When compiling C++ code for versions of Symbian OS earlier than version 9, compilers are explicitly directed to disable C++ exception handling, and any use of the `try`, `catch` or `throw` keywords is flagged as an error.

Symbian OS version 9, by taking advantage of compiler improvements, supports C++ standard exceptions and provides a more open environment. This makes it easier to port existing C++ code onto the Symbian platform.

Symbian OS provides leaves as an alternative to standard C++ exceptions and conventional error checking, which can produce rather awkward code logic. A leave is used to propagate an error to where it can be handled. Leaves are a simple, lightweight exception-handling mechanism which is fundamental to Symbian OS.

What is a leave?

Rather like C++ exceptions, a leave suspends code execution at the point the leave occurs and resumes execution where the leave is trapped. The trap harness in Symbian is a `TRAP` macro – the leave sets the stack pointer to the context of the `TRAP` and jumps to that location, restoring the register values. It does not terminate the flow of execution. The use of `TRAP` is described in more detail in Section 5.4.

The `TRAP` harness and the system function which causes the leave (`User::Leave()`) may be considered analogous to the standard library `setjmp()` and `longjmp()` methods, respectively. A call to `setjmp()` stores information about the location to be "jumped to" in a jump buffer, which is used by `longjmp()` to determine the location to which the point of execution "jumps".

A call to `User::Leave()` or `User::LeaveIfError()` is similar to a C++ `throw` instruction (except for its destruction of stack-based variables, as discussed shortly) while the `TRAP` macros are, in effect, a combination of `try` and `catch`.

However, unlike a C++ `throw`, the leave mechanism simply deallocates objects on the stack – **it does not call any destructors on those objects as it does so**. If a stack object owns a resource which must be deallocated or otherwise "released" as part of destruction, it will leak that resource in the event of a leave, which is clearly unacceptable. This is why Symbian OS has a class-naming convention (described in Chapter 4) which clearly defines whether a class type may be created on the stack. The only classes which may be instantiated and used safely on the stack are T classes, because T classes must not require a destructor (that is they must not contain data which needs to be cleaned up, and are restricted to ownership of built-in types or other T classes). A stack-based T-class object will thus be cleaned up correctly if a leave occurs because, in effect, there is nothing to clean up as the stack unwinds.

R classes may also be created on the stack, but they must be made "leave safe", if used in functions that may leave, by using the cleanup stack as described in Section 5.5.

What causes a leave?

A typical leaving function is one that performs an operation that is not guaranteed to succeed, such as allocation of memory, which may fail under low memory conditions, or creation of a file when there is insufficient disk space.

A leave may occur when a leaving function is called, or by an explicit call to a system function that causes a leave. A function may leave if it:

- Calls code that may leave without surrounding that call with a `TRAP` harness; functions which may leave must be named to indicate the fact, as described in Section 5.2
- Calls one of the system functions that initiates a leave, such as `User::Leave()` or `User::LeaveIfError()` (described below)
- Uses the overloaded form of operator `new` which takes `ELeave` as a parameter (described below).

The overload of operator `new` which takes `ELeave` as a parameter guarantees that the pointer return value will always be valid if a leave has not occurred.

Heap allocation using `new(ELeave)`

Symbian OS overloads the global operator `new` to leave if there is insufficient heap memory for a successful allocation. Use of this overload allows the pointer returned from the allocation to be used without a further test that the allocation was successful (the allocation would leave if it were not). Here is a code fragment which illustrates the use of the operator `new` overload:

```
CCat* InitializeCatL() // The final L is explained in Section 5.2
  {
  CCat* cat = new(ELeave) CCat();
  cat->Initialize();
  return (cat);
  }
```

The code above is preferable to the following code, which requires an additional check to verify that the `cat` pointer has been initialized:

```
CCat* InitializeCat()
  {
  CCat* cat = new CCat();
  if (cat)
    {
    cat->Initialize();
    return (cat);
    }
  else
    return (NULL);
  }
```

Functions in class `User` which cause a leave

- `User::LeaveIfError()` tests an integer parameter passed into it and causes a leave (using the integer value as a leave code) if the value is less than zero, for example, one of the `KErrXXX` error constants defined in `e32std.h`. `User::LeaveIfError()` is useful for turning a non-leaving function which returns a standard Symbian OS error into one which leaves with that value.
- `User::Leave()` doesn't carry out any value checking and simply leaves with the integer value passed into it as a leave code.
- `User::LeaveNoMemory()` also simply leaves, but the leave code is hardcoded to be `KErrNoMemory` which makes it, in effect, the same as calling `User::Leave(KErrNoMemory)`.
- `User::LeaveIfNull()` takes a pointer value and leaves with `KErrNoMemory` if it is `NULL`. It can sometimes be useful, for example, to enclose a call to a non-leaving function which allocates memory and returns a pointer to that memory or `NULL` if it is unsuccessful.

Exam Essentials

- Know that, before v9, Symbian OS does not support standard C++ exceptions (`try/catch/throw`) but uses a lightweight alternative: `TRAP` and leave
- Know that leaves are a fundamental part of Symbian error handling and are used throughout the system
- Understand the similarity between leaves and the `setjmp/longjmp` declarations in C
- Recognize the typical system functions which may cause a leave, including the `User::LeaveXXX()` functions and `new(ELeave)`
- Be able to list typical circumstances which cause a leave (for example, insufficient memory for a heap allocation)
- Understand that `new(ELeave)` guarantees that the pointer return value will always be valid if a leave has not occurred

5.2 How to Work with Leaves

Critical Information

Assume that a function has previously allocated memory on the heap and this memory is referenced only by a local pointer variable. If a leave occurs inside the function, the pointer will be destroyed by the leave (as the stack frame is unwound back to the `TRAP` handler) and the heap memory it references will become unrecoverable, causing a memory leak. The following code illustrates this:

```
void UnsafeFunctionL()
  {
  // Allocates test on the heap
  CTestClass* test = CTestClass::NewL();
  test->FunctionMayLeaveL(); // Unsafe - test may be leaked!
  delete test;
  }
```

The code is unsafe because the memory allocated on the heap in the call to `CTestClass::NewL()` will become inaccessible if the subsequent call to `FunctionMayLeaveL()` does leave. This means that `test` can never be deallocated (it is said to be "orphaned") and will result in a memory leak. The function is not "leave-safe".

To make a function leave-safe, preventing the possibility of memory leaks, any heap objects referenced only by local variables must be pushed onto the cleanup stack before calling any functions which may leave. The cleanup stack, described in Section 5.5, will delete the heap memory should a leave occur.

But how is it possible to know whether a function may leave? Symbian OS has a naming convention in place to indicate this (rather like the C++ `throw(...)` exception specification): if a function may leave, its name must end with a trailing `"L"` to identify it as such. Of all Symbian OS naming conventions, this is probably the most important rule: if a leaving function is not named to indicate its potential to leave, callers of that function may not defend themselves against a leave and may potentially leak memory.

Since it is not part of the C++ standard, the trailing L cannot be checked during compilation, and can sometimes be forgotten, or leaving code introduced to a previously non-leaving function. Symbian OS provides a helpful tool, LeaveScan, that checks code for incorrectly named leaving functions.

The following code shows an example of four possible leaves:

```
TInt UseCat(CCat* aCat); // Forward declaration
CCat* InitializeCatL()
  {
  CCat* cat = new(ELeave) CCat();   // (1)
  CleanupStack::PushL(cat);         // (2) See Section 5.5
  cat->InitializeL();               // (3)
  User::LeaveIfError(UseCat(cat)); // (4)
  CleanupStack::Pop(cat);
  return (cat);
  }
```

Since leaving functions by definition leave with an error code (a "leave code"), they do not also need to return error values. Indeed, any error that occurs in a leaving function should be passed out as a leave; if the function does not leave it is deemed to have succeeded and will return normally. Generally, leaving functions should return void unless they use the return value for a pointer or reference to a resource allocated by the function, as shown above in `InitializeCatL()`.

Exceptions

Member variables are leave-safe

While heap variables referenced only by local variables may be orphaned if a leave occurs, member variables will not suffer a similar fate (unless their destructor neglects to delete them when it is called at some later point). Thus the following code is safe:

```
void CTestClass::SafeFunctionL()
  {
  iMember = CCatClass::NewL(); // Allocates a heap member
  FunctionMayLeaveL();         // Safe for iMember
  }
```

The `CTestClass` object (pointed to by the `this` pointer in `CTestClass::SafeFunctionL()`) is not deleted in the event of a leave. The heap-based `iMember` is stored safely as a pointer member variable, to be deleted at a later stage with the rest of the object, through the class destructor.

When a leave should not occur: constructors and destructors

Neither a constructor nor a destructor should contain code that may leave, since doing so would potentially leak memory.

If a leave occurs in a constructor, it places the object in an indeterminate state. In essence, if a constructor can fail, through lack of the resources necessary to create or initialize the object, it is possible that memory would be leaked. The two-phase construction paradigm must be used to prevent this (see Chapter 6).

Likewise, a leave should never occur in a destructor or in cleanup code. One reason for this is that a destructor could itself be called as part of cleanup following a leave and a further leave at this point would be undesirable, if nothing else because it would mask the initial reason for the leave. More obviously, a leave part-way through a destructor will leave the object destruction incomplete, which may leak its resources.

Exam Essentials

- Know that leaves are indicated by use of a trailing L suffix on functions containing code which may leave (for example, `InitializeL()`)

- Be able to spot functions which are not leave-safe and those which are
- Understand that leaves are used for error handling; code should very rarely both return an error and be able to leave
- Understand the reason why a leave should not occur in a constructor or destructor

5.3 Comparing Leaves and Panics

Critical Information

Leaves occur under exceptional conditions such as out-of-memory or out-of-disk-space, and are also used in place of returning an error. Leaves should only be used to propagate an error or exception to a point in the code which can handle it gracefully. They should not be used to direct the normal flow of program logic. Leaves should always be caught and handled – they do not terminate the flow of execution.

Panics, in contrast, cannot be caught and handled. A panic terminates the thread in which it occurs, and usually the entire application. This results in a poor user experience and, for this reason, panics should only be used in assertion statements to check code logic and fix programming errors during development. (Panics and assertion statements are covered in more detail in Chapter 10.)

If a panic occurs from system or application code during development, it's necessary to find the cause and fix it, since it cannot be handled. Symbian OS panics are documented in the Symbian Developer Library.

Exam Essentials

- Understand the difference between a leave and a panic
- Recognize that panics come about through assertion failures, which should be used to flag programming errors during development
- Recognize that a leave should not be used to direct normal code logic

5.4 What Is a `TRAP`?

Critical Information

Symbian OS provides two trap harness macros, `TRAP` and `TRAPD`, to trap leaves and allow them to be handled. The macros differ only in that `TRAPD` declares the variable in which the leave error code is returned, while code using `TRAP` must declare a `TInt` variable itself first. Thus the following code segment:

```
TRAPD(result, MayLeaveL());
if (KErrNone!=result)
  {
  // Handle error
  }
```

is equivalent to:

```
TInt result;
TRAP(result, MayLeaveL());
if (KErrNone!=result)
  {
  // Handle error
  }
```

If a leave occurs inside the `MayLeaveL()` function, which is executed inside the harness, the program control will return immediately to the `TRAP` harness macro. The variable `result` will contain the error code associated with the leave (that is that passed as a parameter to the `User::Leave()` system function) or will be `KErrNone` if no leave occurred.

Any functions called by `MayLeaveL()` are executed within the `TRAP` harness, and so on recursively, and any leave that occurs during the execution of `MayLeaveL()` is trapped, returning the error code into `result`. Alternatively, `TRAP` macros can be nested to catch and handle leaves at different levels of the code, where they can best be dealt with.

Each `TRAP` has an impact on executable size and execution speed. Both entry to and exit from a `TRAP` macro results in kernel executive calls (`TTrap::Trap()` and `TTrap::UnTrap()`) which switch the user-side code into processor-privileged mode in order to access kernel resources. These are quite expensive in terms of execution speed. In addition, a structure is allocated at run-time to hold the current contents of the thread's stack in order to return to that state should a leave occur. These factors, combined with the inline code generated by the `TRAP` macro, add up to a fairly significant overhead.

The number of `TRAP`s should be minimized where possible, and code that uses the `TRAP` macro several times in one function, or nests a series of them, should be inspected to determine if it can be refactored. For example, the following function must not leave but needs to call a number of functions that do. At first sight, it might seem straightforward enough simply to put each call in a `TRAP` harness.

```
TInt MyNonLeavingFunction()
  {
  TRAPD(result, FunctionMayLeaveL());
  if (KErrNone==result)
    TRAP(result, AnotherFunctionWhichMayLeaveL());
```

```
  if (KErrNone==result)
    TRAP(result, PotentialLeaverL());
  // Handle any error if necessary
  return (result);
  }
```

It can, however, be refactored as follows, for efficiency.

```
TInt MyNonLeavingFunction()
  {
  TRAPD(result, MyNewLeavingFunctionL());
  // Handle any error if necessary
  return (result);
  }
void MyNewLeavingFunctionL()
  {
  FunctionMayLeaveL();
  AnotherFunctionWhichMayLeaveL();
  PotentialLeaverL();
  }
```

Of course, code is rarely as trivial as this example and where a number of leaving calls are packaged together, as in `MyNewLeavingFunctionL()` above, if a leave occurs it will not be clear which function actually left.

Symbian OS programs have at least one `TRAP`, if only at the topmost level to catch any leaves that are not trapped elsewhere (the application framework provides a `TRAP` for UI applications which is not visible).

Exam Essentials

- Recognize the characteristics of a `TRAP` handler
- Understand that, for efficiency, use of `TRAP`s should be kept to a minimum

5.5 The Cleanup Stack

Critical Information

Section 5.2 explained that a memory leak can occur as a result of a leave when there are heap objects accessible only through pointers local to the function that leaves. This is because the leave will not call `delete` on the pointer; it will instead be destroyed without freeing the heap memory it references, which "orphans" that memory and causes a leak. In effect, the code switches back to (a copy of) the stack frame from the time the `TRAP` harness was called, when the local pointer variable wasn't defined, and thus a memory leak occurs.

This means that C-class objects, which are always created on the heap as described in Chapter 4, are not leave-safe unless they are otherwise accessible for safe destruction (for example, as member variables of an object which is destroyed regardless of the leave). R-class objects are generally not leave-safe either, since the resources they own must be freed in the event of a leave (through a call to the appropriate `Close()` or `Release()` function). If this call cannot made by an object that exists at the time the `TRAP` harness is entered, the resource is orphaned.

The following code creates an object of a C class (`CCat`) on the heap, referenced only by an automatic variable, `cat`. After creating the object, a function which may potentially leave, `InitializeL()`, is called. The heap-based object is not leave-safe and neither the heap memory it occupies nor any objects it owns would be destroyed if `InitializeL()` left.

```
void UnsafeFunctionL()
  {
  CCat* cat = new(ELeave) CCat();
  cat->InitializeL(); // Potential leaving function orphans cat
  DoSomethingElseL() // May also leave
  delete cat;
  }
```

One way to resolve this would be to place a `TRAP` (or `TRAPD`) macro around the call to `InitializeL()` to catch any leaves. However, as Section 5.4 explained, the use of `TRAP`s should be limited, where possible, to optimize the size and run-time speed of the compiled binary.

Introducing the cleanup stack

The solution is to use the cleanup stack, which is accessed through the static member functions of class `CleanupStack`, defined in `e32base.h`:

```
class CleanupStack
  {
public:
  IMPORT_C static void PushL(TAny* aPtr);
  IMPORT_C static void PushL(CBase* aPtr);
  IMPORT_C static void PushL(TCleanupItem anItem);
  IMPORT_C static void Pop();
  IMPORT_C static void Pop(TInt aCount);
  IMPORT_C static void PopAndDestroy();
  IMPORT_C static void PopAndDestroy(TInt aCount);
  IMPORT_C static void Check(TAny* aExpectedItem);
  inline static void Pop(TAny* aExpectedItem);
  inline static void Pop(TInt aCount,TAny* aLastExpectedItem);
  inline static void PopAndDestroy(TAny* aExpectedItem);
  inline static void PopAndDestroy(TInt aCount,TAny* aLastExpectedItem);
  };
```

Pointers to objects that are not otherwise leave-safe should be placed on the cleanup stack before calling code that may leave. This ensures they are destroyed correctly if a leave occurs; in the event of a leave, the cleanup stack manages the deallocation of all objects which have been placed upon it.

The following code illustrates a leave-safe version of `UnsafeFunctionL()` above:

```
void SafeFunctionL()
  {
  CCat* cat = new(ELeave) CCat;
  // Push onto the cleanup stack before calling a leaving function
  CleanupStack::PushL(cat);
  cat->InitializeL();
  DoSomethingElseL()     // May also leave
  // Pop from cleanup stack
  CleanupStack::Pop(cat);
  delete cat;
  }
```

If no leaves occur, the `CCat` pointer is popped from the cleanup stack and the object to which it points is deleted. This code could equally well be replaced by a single call to `CleanupStack::PopAndDestroy (cat)` which pops the pointer and makes a call to the destructor in one step. If `InitializeL()` or `DoSomethingElseL()` leaves, the `cat` object is destroyed by the cleanup stack itself as part of leave processing, as will be discussed shortly. The cleanup stack is rather like a Symbian OS version of the standard C++ library's smart pointer, `auto_ptr`.

How to use the cleanup stack

Pointers are pushed onto and popped off the cleanup stack in strict order: since it's a stack, a series of `Pop()` calls must occur in the reverse order of the `PushL()` calls. It is possible to `Pop()` or `PopAndDestroy()` one or more objects without naming them, but it's a good idea to name the object popping off. In debug builds, the cleanup stack will check that the item it is popping off is the same as the one passed in, and panic if not, to indicate that the cleanup stack is not in the expected state.

Checking averts any potential cleanup stack "imbalance" bugs which can occur when `Pop()` removes an object from the cleanup stack which was not the one intended.

```
void ContrivedExampleL()
  {// Note that each object is pushed onto the cleanup stack
  // immediately it is allocated, in case the succeeding allocation
  // leaves.
  CSiamese* sealPoint = CSiamese::NewL(ESeal);
  CleanupStack::PushL(sealPoint);
```

```
CSiamese* chocolatePoint = CSiamese::NewL(EChocolate);
CleanupStack::PushL(chocolatePoint);
CSiamese* violetPoint = CSiamese::NewL(EViolet);
CleanupStack::PushL(violetPoint);
CSiamese* bluePoint = CSiamese::NewL(EBlue);
CleanupStack::PushL(bluePoint);
sealPoint->CatchMouseL();
// Other leaving function calls, some of which use the cleanup stack
...
// Various ways to remove the objects from the stack and delete them:
// (1) All with one anonymous call - OK
// CleanupStack::PopAndDestroy(4);
// (2) Each object individually to verify the code logic
// Note the reverse order of Pop() to PushL()
// This is quite long-winded and probably unnecessary in this
// example
// CleanupStack::PopAndDestroy(bluePoint);
// CleanupStack::PopAndDestroy(violetPoint);
// CleanupStack::PopAndDestroy(chocolatePoint);
// CleanupStack::PopAndDestroy(sealPoint);
// (3) All at once, naming the last object - best solution
CleanupStack::PopAndDestroy(4, sealPoint);
}
```

Why does `CleanupStack::PushL()` leave?

`PushL()` is a leaving function because it may need to allocate memory for pointer storage and thus may fail in low-memory situations. However, the object passed into the `PushL()` method will not be orphaned if `PushL()` does leave. This is because, when the cleanup stack is created, it has at least one spare slot. When `PushL()` is called, the pointer is added to the next vacant slot, and then, if there are no remaining slots available, the cleanup stack implementation attempts to allocate more slots for future usage. If this allocation fails, only then does a leave occur; however, the pointer passed in has already been stored safely on the cleanup stack and the object it refers to will be safely cleaned up.

For efficiency, the cleanup stack is expanded four slots at a time. In addition, `Pop()` does not release slots when pointers are popped out of them, so a `PushL()` call frequently does not need to make any further allocation and can be guaranteed to succeed.

When to remove an item from the cleanup stack

It should never be possible for an object to be cleaned up more than once. If a pointer to an object on the cleanup stack is later stored elsewhere, say as a member variable of another object which is accessible after a leave, the pointer should then be popped from the cleanup stack. If the pointer were retained on the cleanup stack, cleanup after a leave would destroy it, but the other object storing the pointer would also be likely to do so, usually in its own destructor. An attempt to delete an object which has already been released back to the heap will cause a system panic. To avoid this, objects should be referred to either by another object or by the cleanup stack, but not by both.

```
void TransferOwnershipExampleL
  {
  CCat* cat = new(ELeave) CCat();
  CleanupStack::PushL(cat); // The next function may leave
  iMemberObject->TakeOwnershipL(cat);// iMemberObject owns it now
  CleanupStack::Pop(cat);   // remove from cleanup stack, don't delete
  }
```

This is the reason that pointers which are class member variables (prefixed by `i`) should not be pushed onto the cleanup stack. The object may be accessed through the owning object which destroys it when appropriate, typically in its destructor, so does not need to be made leave-safe through use of the cleanup stack.

No panic occurs if objects are pushed to, or popped from, the cleanup stack more than once. The problem occurs if the cleanup stack tries to delete the object twice, either through multiple calls to `PopAndDestroy()` or in the event of a leave.

When to leave a pointer on the cleanup stack

In a function, if a pointer to an object is pushed onto the cleanup stack and remains on it when that function returns, the Symbian OS convention is to append a `C` to the function name. This indicates to the caller that, when the function returns successfully, the cleanup stack has additional pointers on it.

This approach is typically used by `CBase`-derived classes which define static functions to instantiate an instance of the class and leave a pointer to it on the cleanup stack, as described in Chapter 6. The following code creates an object of type `CSiamese` (as used in an earlier example) and leaves a pointer to it on the cleanup stack. This function is useful because the caller can instantiate `CSiamese` and immediately call a leaving function without itself needing to push the pointer onto the cleanup stack:

```
/*static*/ CSiamese* CSiamese::NewLC(TPointColor aPointColour)
  {
  CSiamese* me = new(ELeave) CSiamese(aPointColour);
  CleanupStack::PushL(me); // Make this leave-safe...
  me->ConstructL();
  return (me); // me remains on the cleanup stack
  }
```

However, functions that leave objects on the cleanup stack must not be called from immediately inside a `TRAP` harness. If objects are pushed onto the cleanup stack inside a TRAP and a leave does not occur, they must be popped off again before exiting the macro, otherwise a panic occurs. This is because the cleanup stack stores objects in nested levels; each level is confined within a TRAP, and must be empty when the code inside it returns. Thus, the following code panics with E32USER-CBASE 71 when it returns to the TRAPD macro.

```
CSiamese* MakeSiamese(TPointColor aPointColour)
  {// The next line will cause a panic (E32User-CBase 71)
  CSiamese* pCat = TRAPD(r, CSiamese::NewLC(aPointColour));
  return (pCat);
  }
```

The D suffix naming convention

Section 5.4 introduced the use of a suffixed D (on the TRAPD macro) to indicate that a harness macro declares an integer variable to hold the leave code of the leaving function it TRAPs.

However, there is an additional meaning for the use of a D suffix, when used on function names: a function whose name ends in D will take responsibility for destroying the object on which it is called. Since the function will delete the object when it is finished with it, any calling code should not attempt to do so. A good example of such a function is CEikDialog::ExecuteLD().

Using the cleanup stack with T, R and M classes

Up to this point, the discussion has only really referred to pushing pointers to CBase-derived objects (objects of C classes, as discussed in Chapter 4) onto the cleanup stack. However, there are three overloads of the PushL() method used to place items onto the cleanup stack. The overloads determine how the item is later destroyed when it is cleaned up (either when a leave occurs or through a call to Cleanup-Stack::PopAndDestroy()).

```
IMPORT_C static void PushL(CBase* aPtr);
```

When the first overload, which takes a pointer to a CBase-derived object, is used to push a pointer onto the cleanup stack, it will be configured to be destroyed by invoking delete on the pointer, thus calling the virtual destructor of the CBase-derived object. This is the reason that the CBase class has a virtual destructor – so that C-class objects can be placed on the cleanup stack and destroyed safely if a leave occurs.

```
IMPORT_C static void PushL(TAny* aPtr);
```

If a class with the characteristics of a Symbian OS Class is defined but does not derive from CBase, the second overload of PushL(), CleanupStack::PushL(TAny*), will be invoked to push a pointer to such an object onto the cleanup stack. This overload means that, if the cleanup stack later comes to destroy the object, its heap memory is simply deallocated (by invoking User::Free()) and delete is not called on it (and hence no destructor is called).

Indeed, the CleanupStack::PushL(TAny*) overload is used whenever any heap-based object which does not derive from CBase is pushed onto the cleanup stack. This includes, for example, T-class objects and structs which have been allocated, for some reason, on the heap. T classes do not have destructors, and thus have no requirement for cleanup beyond deallocation of the heap memory they occupy, so will be cleaned up correctly.

When heap descriptor objects (of class HBufC, described in Section 7.2) are pushed onto the cleanup stack, the CleanupStack::PushL (TAny*) overload is also used. This is because HBufC objects are, in effect, heap-based objects of a T class. Objects of type HBufC require no destructor invocation – the only cleanup necessary is that required to free the heap cells for the descriptor object.

```
IMPORT_C static void PushL(TCleanupItem anItem);
```

The third overload, CleanupStack::PushL(TCleanupItem), takes an object of type TCleanupItem and is designed to allow objects with types of cleanup processing other than CBase deletion or simple deallocation to take advantage of the cleanup stack. This means that R or M classes, or classes with customized cleanup routines, can also be made leave-safe.

A TCleanupItem object encapsulates a pointer to the object to be stored on the cleanup stack and a pointer to a function that provides cleanup for that object. The cleanup function can be a local function or a static method of a class. A leave or a call to PopAndDestroy() removes the object from the cleanup stack and calls the cleanup function provided by the TCleanupItem.

Symbian OS also provides a set of template utility functions, each of which generates an object of type TCleanupItem and pushes it onto the cleanup stack:

- CleanupReleasePushL() – the cleanup method calls Release() on the object in question. This method is typically used to make leave-safe an object referenced through an M-class (interface) pointer.
- CleanupDeletePushL() – the cleanup method calls delete on the pointer passed into the function. This is also typically used

for M-class objects, which should not be pushed onto the cleanup stack using the `CleanupStack::PushL(TAny*)` overload, since an M-class pointer cannot simply be deallocated by a call to `User::Free()`.

- `CleanupClosePushL()` – the cleanup method calls `Close()` on the object in question. This method is typically used to make stack-based R-class objects leave-safe. For example:

```
void UseFilesystemL()
  {
  RFs theFs;
  User::LeaveIfError(theFs.Connect());
  CleanupClosePushL(theFs);
  ... // Call functions which may leave
  CleanupStack::PopAndDestroy(&theFs);
  }
```

- `CleanupArrayDeletePushL()` – this method is used to push a pointer to a heap-based C++ array of T-class objects (or built-in types) on to the cleanup stack. When `PopAndDestroy()` is called, the memory allocated for the array is cleaned up using `delete[]`. No destructor is called on the elements of the array.

```
TTest* pTestArray = new(ELeave) TTest[KMaxArraySize];
CleanupArrayDeletePushL(pTestArray);
... // Call functions which may leave
CleanupStack::PopAndDestroy(pTestArray); // Calls delete[] pTestArray
```

Creating the cleanup stack

The cleanup stack is created as follows and, once created, any leaving code which uses it must be called within a base-level `TRAP` harness:

```
CTrapCleanup* theCleanupStack = CTrapCleanup::New();
... // Code that uses the cleanup stack within a TRAP macro
delete theCleanupStack;
```

It isn't necessary to create a cleanup stack for a GUI application, since the application framework creates one. However, a cleanup stack must be created if writing a server, a simple console-test application or any code which creates an additional thread which uses the cleanup stack (or calls code which does so).

Exam Essentials

- Know how to use the cleanup stack to make code leave-safe, so memory is not leaked in the event of a leave

- Understand that `CleanupStack::PushL()` will not leak memory even if it leaves
- Know the order in which to remove items from the cleanup stack, and how to use `CleanupStack::PopAndDestroy()` and `CleanupStack::Pop()`
- Recognize correct and incorrect use of the cleanup stack
- Understand the consequences of putting a C class on the cleanup stack if it does not derive from `CBase`
- Know how to use `CleanupStack::PushL()` and `CleanupXXXPushL()` for objects of C, R, M and T classes and `CleanupArrayDeletePushL()` for C++ arrays
- Understand the meaning of the Symbian OS function suffixes C and D

5.6 Detecting Memory Leaks

Critical Information

Memory is a limited resource on Symbian OS and must be managed carefully to ensure it is not wasted by memory leaks. In addition, an application must gracefully handle any exceptional conditions arising when memory resources are exhausted. Symbian OS provides a set of debug-only macros that can be added directly to code to check that memory is not leaked.

There are a number of macros available, but the most commonly used are defined as follows:

```
#define __UHEAP_MARK User::__DbgMarkStart(RHeap::EUser)
#define __UHEAP_MARKEND User::__DbgMarkEnd(RHeap::EUser,0)
```

The `__UHEAP_MARK` and `__UHEAP_MARKEND` macros verify that the default user heap is consistent. The check is started by using `__UHEAP_MARK` and a subsequent call to `__UHEAP_MARKEND` performs the verification. If any heap cells were allocated after the call to `__UHEAP_MARK` and then not freed back to the heap before the call to `__UHEAP_MARKEND`, a panic will occur in debug builds to indicate that the heap is inconsistent. The panic raised is `ALLOC nnnnnnnn`, where `nnnnnnnn` is a hexadecimal pointer to the first orphaned heap cell.

The heap-checking macros can be nested inside each other and used anywhere in code. They are ignored by release builds of Symbian OS, so they can be left in production code without any impact on the code size or speed.

Exam Essentials

- Recognize the use of the `__UHEAP_MARK` and `__UHEAP_MARKEND` macros to detect memory leaks

References

[Babin 2005 Chapter 4]
[Harrison 2004 Chapter 1]
[Stichbury 2004 Chapters 2, 3 and 17]

6

Two-Phase Construction and Object Destruction

Introduction

Symbian OS takes memory efficiency very seriously indeed, because it was designed to perform well on devices with limited memory. As Chapter 5 describes, it uses memory-management models such as the cleanup stack to ensure that memory is not leaked, even under error conditions or in exceptional circumstances, such as when there is insufficient free memory to complete an allocation. Two-phase construction is an idiom used extensively in Symbian OS code to provide a means by which heap-based objects may be constructed fully initialized, even when that initialization code may leave.

6.1 Two-Phase Construction

Critical Information

The following line of code allocates an object of type `CExample` on the heap and sets the value of `foo` accordingly:

```
CExample* foo = new CExample();
```

The code calls the `new` operator, which allocates a `CExample` object on the heap if there is sufficient memory available. Having done so, it calls the constructor of class `CExample` to initialize the object.

As Chapter 5 describes, the cleanup stack can be used to ensure that the `CExample` object is correctly cleaned up in the event of a leave. But it is not possible to do so inside the `new` operator between allocation of the object and invocation of its constructor. If the `CExample` constructor itself leaves, the memory already allocated for the object and any additional memory the constructor may have allocated will be orphaned.

This gives rise to a key rule of Symbian OS memory management: **no code within a C++ constructor should ever leave**.

Of course, to fully initialize an object, it may be necessary to write code that leaves, say to allocate memory to store another object, or to allocate a descriptor into which is read a configuration file, which may be missing or corrupt. There are many reasons why initialization may fail, and it is under these circumstances that two-phase construction is necessary when instantiating a fully initialized object.

Two-phase construction breaks object construction into two parts, or phases:

1. A basic constructor which cannot leave.
   It is this constructor which is called by the `new` operator. It implicitly calls base-class constructors and may also invoke functions that cannot leave and/or initialize member variables with default values or those supplied as arguments to the constructor.

2. A class method (typically called `ConstructL()`).
   This method may be called separately once the object, allocated and constructed by the `new` operator, has been pushed onto the cleanup stack; it will complete construction of the object and may safely perform operations that may leave. If a leave does occur, the cleanup stack calls the destructor to free any resources which have already been successfully allocated and destroys the memory allocated for the object itself.

```
class CExample : public CBase
  {
public:
  CExample();  // Guaranteed not to leave
  ~CExample(); // Must cope with partially constructed objects
  void ConstructL(); // Second-phase construction code - may leave
  ... // Omitted for clarity
  };
```

This is the simplest implementation of two-phase construction, and expects the calling code instantiating an object of the class to call the second-phase construction function. However, member functions in the class will not be able rely on the object having being fully constructed. Some callers may forget to call `ConstructL()` after instantiating the object and, at the least, will find it a burden since it's not a standard C++ requirement to do so.

For this reason, it is preferable to make the call to the second-phase construction function within the class itself. Obviously the code cannot do this from within the simple constructor, since this takes it back to the original problem of calling a method which may leave.

A commonly used pattern in Symbian OS code is to provide a static function which wraps both phases of construction, providing a simple and easily identifiable means to instantiate objects of a class on the heap. The function is typically called `NewL()`. A `NewLC()` function is often provided too; this is identical except that it leaves the constructed object on the cleanup stack for convenience.

```
class CExample : public CBase
  {
public:
  static CExample* NewL();
  static CExample* NewLC();
  ~CExample(); // Must cope with partially constructed objects
  ... // Other public methods, eg Foo(), Bar()
private:
  CExample();         // Guaranteed not to leave
  void ConstructL(); // Second-phase construction code, may leave
  ...                 // Omitted for clarity
  };
```

The `NewL()` function is static so that it can be called without first having an existing instance of the class. The non-leaving constructors and second-phase `ConstructL()` functions have been made private so that a caller cannot instantiate objects of the class except through `NewL()` and to prevent calls to `ConstructL()` after the object has been fully constructed.

The use of the `NewL()` function for two-phase construction prevents all of the following incorrect object constructions:

```
CExample example; // BAD! C classes should not be created on the stack

CExample* example = new CExample(); // Caller must test for success

if (example)
    {// call a method on example
    example->Foo(); // ConstructL() for example has not been called
    }

CExample* example = new(ELeave) CExample();
example->Foo(); // ConstructL() wasn't called, example is not
                // fully constructed
```

Typical implementations of `NewL()` and `NewLC()` are as follows:

```
CExample* CExample::NewLC()
  {
  CExample* me = new (ELeave) CExample(); // First-phase construction
  CleanupStack::PushL(me);
```

```
  me->ConstructL(); // Second-phase construction
  return (me);
  }
CExample* CExample::NewL()
  {
  CExample* me = CExample::NewLC();
  CleanupStack::Pop(me);
  return (me);
  }
```

The NewL() function is implemented in terms of the NewLC() function rather than the other way around (which would be slightly less efficient since this would require an extra PushL() call on the cleanup stack). Note the use of the Symbian OS overload of operator new(ELeave), as discussed in Chapter 5. This implementation will leave if it fails to allocate the required amount of memory. This means that there is no need to check for a NULL pointer after a call to new(ELeave).

Each function returns a fully constructed object, or will leave either if there is insufficient memory to allocate the object (that is, if operator new(ELeave) leaves) or if the second-phase ConstructL() function leaves for any reason. If second-phase construction does fail, the cleanup stack ensures both that the partially constructed object is destroyed and that the memory it occupies, allocated in the first phase, is returned to the heap.

The NewL() and NewLC() functions may, of course, take parameters with which to initialize the object. These may be passed to the simple constructor in the first phase or to the second-phase ConstructL() function, or to both.

Deriving from a class which uses two-phase construction

If any class is to be subclassed, the default constructor should be made protected rather than private so that the compiler can construct the derived classes. C++ will ensure that the first-phase constructor of a base class is called prior to calling the derived-class constructor, in the first phase.

However, two-phase construction is not part of standard C++ construction, so the second-phase constructor of a base class will not be called automatically when constructing a derived class. The second-phase ConstructL() method should be made private if it should **not** be called by a derived class, or protected if it does need to be called (and this should be clearly documented).

Often, if a class is intended for extension through inheritance and uses the two-phase construction pattern, it supplies a protected method called BaseConstructL() rather than ConstructL(). The derived class calls this method in its own ConstructL() method to ensure that the base-class object is fully initialized.

Two-phase construction and Symbian OS types

Two-phase construction is typically used only for C classes, since T classes do not usually require complex construction code (because they do not contain heap-based member data) and R classes are usually created uninitialized, requiring their callers to call `Connect()` or `Open()` to associate the R-class object with a particular resource. Chapter 4 discusses the characteristics of the various Symbian OS classes in more detail.

Exam Essentials

- Know why code should not leave inside a constructor
- Recognize that two-phase construction is used to avoid the accidental creation of objects with undefined state
- Understand that constructors and second-phase `ConstructL()` methods are given private or protected access specifiers in classes which use two-phase construction, to prevent their inadvertent use
- Understand how to implement two-phase construction, and how to construct an object which derives from a base class which also uses a two-phase method of initialization
- Know the Symbian OS types (C classes) which typically use two-phase construction

6.2 Object Destruction

Critical Information

When implementing the standard Symbian OS two-phase construction idiom, it is important to consider the destructor code carefully. A destructor must be coded to release all the resources that an object owns. However, the destructor may be called to clean up partially constructed objects if a leave occurs in the second-phase `ConstructL()` function. The destructor code cannot assume that the object is fully initialized and should beware of calling functions on pointers which may not yet be set to point to valid objects.

The memory for a `CBase`-derived object is guaranteed to be set to binary zeroes on first construction (see Section 4.3). It is safe for a destructor to call `delete` on a `NULL` pointer, but the destructor code should beware of attempting to free other resources without checking whether the handle or pointer which refers to them is valid, for example:

```
CExample::~CExample()
  {
  if (iMyAllocatedMember)    // iMyAllocatedMember may be NULL
```

```
    {
    iMyAllocatedMember->DoSomeCleanupPreDestruction();
    delete iMyAllocatedMember;   // No need to set it to NULL
    }
}
```

Exam Essentials

- Know that it is neither efficient nor necessary to set a pointer to `NULL` after deleting it in destructor code
- Understand that a destructor must check before dereferencing a pointer in case it is `NULL`, but need not check if simply calling `delete` on that pointer

References

[Babin 2005 Chapter 4]
[Harrison 2004 Chapter 1]
[Stichbury 2004 Chapter 4]

7

Descriptors

Introduction

A Symbian OS string is known as a *descriptor*, because it is self-describing. A descriptor holds the length of the string of data it represents as well as its type, which identifies the underlying memory layout of the descriptor data.

Descriptors have something of a reputation among Symbian OS programmers, because they take some time to get used to. The key point to remember is that they were designed to be very efficient on low-memory devices, using the minimum amount of memory necessary to store the string, while describing it fully in terms of its length and layout.

7.1 Features of Symbian OS Descriptors

Critical Information

Descriptors are not like standard C++ strings, Java strings or the MFC `CString` (to take just three examples), because their underlying memory allocation and cleanup must be managed by the programmer. But descriptors are also not like C strings: they protect against buffer overrun and don't rely on `NULL` terminators to determine the length of the string.

Character size

The descriptors in early releases of Symbian OS, up to and including Symbian OS v5, had 8-bit native characters. Since that release, Symbian OS has been built to support Unicode character sets with wide (16-bit) characters by default. The character width of descriptor classes can be identified from their names. If the class name ends in 8 (for example

`TPtr8`) it has narrow (8-bit) characters, while a descriptor class name ending with 16 (for example `TPtr16`) manipulates 16-bit character strings.

There is also a set of neutral classes which have no number in their name (for example `TPtr`). The neutral classes are `typedef`'d to the character width set by the platform. On all releases of Symbian OS since v5u (first used in the Ericsson R380 mobile phone), the neutral classes are implicitly wide strings. The neutral classes were defined for source compatibility purposes to ease the switch between narrow and wide builds and, although today Symbian OS is always built with wide characters, it is recommended practice to use the neutral descriptor classes where the character width does not need to be stated explicitly.

Descriptor data

Descriptors are strings and can contain text data. However, they can also be used to manipulate binary data, because they don't rely on a `NULL` terminating character to determine their length, since it is instead built into the descriptor object. To work with binary data, the 8-bit descriptor classes should be used explicitly.

The unification of binary and string-handling APIs makes their use easier for programmers – for example, the APIs to read from and write to a file all take an 8-bit descriptor, regardless of whether the file contains human-readable strings or binary data.

Memory management

The descriptor classes do not dynamically manage the memory used to store their data. The modification methods check that the maximum length of the descriptor is sufficient for the operation to succeed. If it is not, they do not re-allocate memory for the operation but instead panic (see Chapter 10) to indicate that an overflow would occur if they proceeded. In the event of such a panic it can be assumed that no descriptor data was modified.

Thus, before calling a descriptor method which expands the data, it is necessary to check that there is sufficient memory available for it to succeed. Of course, the resulting length of the descriptor can be less than the maximum length allowed (the remainder is left unused) and the contents of the descriptor can shrink and expand up to the maximum length allocated to the descriptor.

Exam Essentials

- Understand that Symbian OS descriptors may contain text or binary data

- Know that descriptors may be narrow (8-bit), wide (16-bit) or neutral (which is 16-bit since Symbian OS is built for Unicode)
- Understand that descriptors do not dynamically extend the data area they reference, so will panic if too small to store data resulting from a method call

7.2 The Symbian OS Descriptor Classes

Critical Information

`TDesC`

Apart from the literal descriptors discussed in Section 7.8, all the descriptor classes derive from the base class `TDesC`. The T prefix indicates a simple type class (see Chapter 4 for Symbian OS class-naming conventions), while the C suffix reflects that the class defines a *non-modifiable* type of descriptor, one whose contents are constant.

As the base class, `TDesC` defines the fundamental layout of every descriptor type: the first 4 bytes always hold the length of the data the descriptor currently contains. In fact, only 28 of the available 32 bits are used to hold the length of the descriptor data (the other 4 bits are reserved for another purpose, to be discussed shortly). This means that the maximum length of a descriptor is limited to 2^{28} bytes (256 MB) which should be more than sufficient! The `Length()` method in `TDesC` returns the length of the descriptor. This method is never overridden by its subclasses since it is equally valid for all types of descriptor.

However, access to the descriptor data is different depending on the implementation of the derived descriptor classes (buffer or pointer, see Section 7.3). To identify each of the derived classes, the top 4 bits of the 4 bytes that store the length of the descriptor object are used to indicate the type of descriptor. When a descriptor operation needs the correct address in memory for the beginning of the descriptor data, it uses the `Ptr()` method of the base class, `TDesC`, which looks up the type of descriptor and returns the correct address for the beginning of the data.

`TDesC` provides methods for determining the length of the descriptor and accessing the data. Using these `Length()` and `Ptr()` methods respectively, the `TDesC` base class can implement all the operations typically possible on a constant string object, such as data access, comparison and search (all of which are documented in full in the Symbian Library). The derived classes all inherit these methods, and in consequence all constant descriptor manipulation is implemented by `TDesC`, regardless of the type of the descriptor.

`TDes`

The modifiable descriptor types all derive from the base class `TDes`, which is itself a subclass of `TDesC`. `TDes` has an additional member variable to store the maximum length of data allowed for the current memory allocated to the descriptor. The `MaxLength()` method of `TDes` returns this value. Like the `Length()` method of `TDesC`, it is not overridden by derived classes.

`TDes` defines a range of methods to manipulate modifiable string data, including those to append, fill and format the descriptor data. Again, all the manipulation code is implemented by `TDes` and inherited by the derived classes.

Derived descriptor classes

Descriptors come in two basic layouts: pointer descriptors, in which the descriptor holds a pointer to the location of a character string stored elsewhere; and buffer descriptors, where the string of characters forms part of the descriptor.

Pointer descriptors: `TPtrC` and `TPtr`

The string data of a pointer descriptor is separate from the descriptor object itself and can be stored in ROM, on the heap or on the stack. The memory that holds the data is not "owned" by the descriptor and pointer descriptors are agnostic about where the memory they point to is actually stored. The pointer descriptors themselves are usually stack-based, but they can be used on the heap, for example as a member variable of a `CBase`-derived class.

In a non-modifiable pointer descriptor (`TPtrC`), the pointer to the data follows the length word, thus the total size of the descriptor object is two words. `TPtrC` is the equivalent of using `const char*` when handling strings in C. The data can be accessed but not modified: that is, the data in the descriptor is constant. All the non-modifying operations defined in the `TDesC` base class are accessible to objects of type `TPtrC`. The class also defines a range of constructors to allow `TPtrC` to be constructed from another descriptor, a pointer into memory, or a zero-terminated C string.

```
// Literal descriptors are described later in this chapter
_LIT(KDes, "Sixty zippers were quickly picked from the woven jute bag");
TPtrC pangramPtr(KDes); // Constructed from a literal descriptor
TPtrC copyPtr(pangramPtr);      // Copy constructed from another TPtrC
TBufC<100> constBuffer(KDes); // Constant buffer descriptor
TPtrC ptr(constBuffer);                 // Constructed from a TBufC
// TText8 is a single (8-bit) character, equivalent to unsigned char
const TText8* cString = (TText8*)"Waltz, bad nymph, for quick jigs vex";
// Constructed from a zero-terminated C string
TPtrC8 anotherPtr(cString);
TUint8* memoryLocation; // Pointer into memory initialized elsewhere
TInt length;            // Length of memory to be represented
...
TPtrC8 memPtr(memoryLocation,length);
```

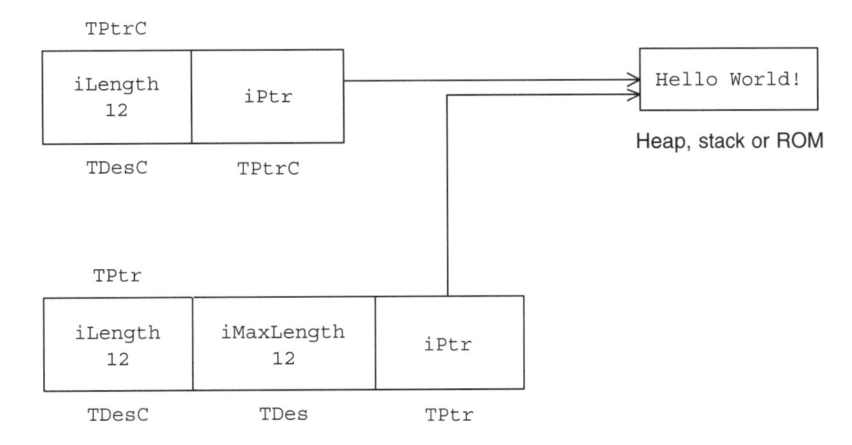

Figure 7.1 Memory layout of pointer descriptors

In a modifiable pointer descriptor (`TPtr`), the data location pointer (4 bytes) follows the maximum length of `TDes` (4 bytes), which itself follows the length word of `TDesC` (4 bytes); the descriptor object is three words in total. Figure 7.1 compares the memory layouts of `TPtr` and `TPtrC`.

The `TPtr` class can be used for access to and modification of a character string or binary data. All the modifiable and non-modifiable base-class operations of `TDes` and `TDesC`, respectively, may be performed on a `TPtr`. The class defines constructors to allow objects of type `TPtr` to be constructed from a pointer into an address in memory, setting the length and maximum length as appropriate.

The compiler also generates implicit default and copy constructors, since they are not explicitly declared protected or private in the class. A `TPtr` object may be copy-constructed from another modifiable pointer descriptor, for example, by calling the `Des()` method on a non-modifiable buffer (see Section 7.4).

```
_LIT(KLiteralDes1, "Jackdaws love my big sphinx of quartz");
TBufC<60> buf(KLiteralDes1); // TBufC is described later
TPtr ptr(buf.Des()); // Copy construction - can modify the data in buf
TInt length = ptr.Length();        // Length=37 characters
TInt maxLength = ptr.MaxLength(); // Maximum length=60 chars, as for buf
TUint8* memoryLocation; // Valid pointer into memory
...
TInt len = 12;            // Length of data to be represented
TInt maxLen = 32;         // Maximum length to be represented
// Construct a pointer descriptor from a pointer into memory
TPtr8 memPtr(memoryLocation, maxLen);        // length=0, max=32
TPtr8 memPtr2(memoryLocation, len, maxLen); // length=12, max=32
```

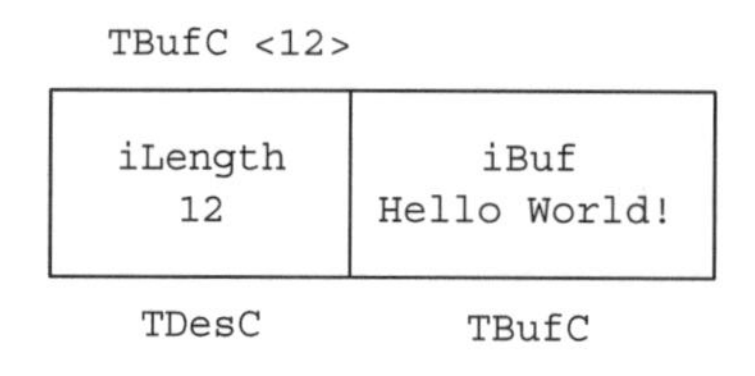

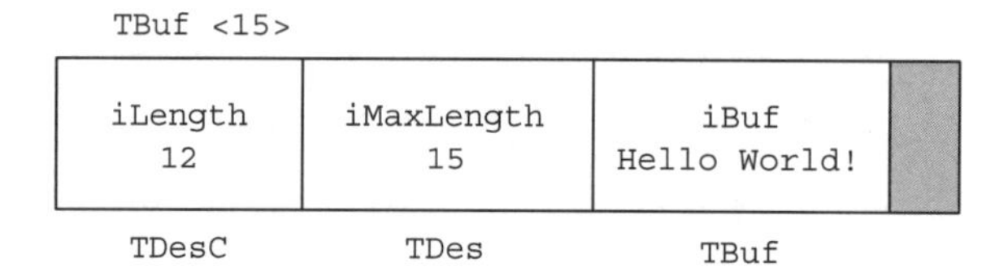

Figure 7.2 Memory layout of buffer descriptors

Stack-based buffer descriptors **`TBufC`** *and* **`TBuf`**

The stack-based buffer descriptors may be modifiable or non-modifiable. The string data forms part of the descriptor object, located after the length word in a non-modifiable descriptor and after the maximum-length word in a modifiable buffer descriptor. Figure 7.2 compares the memory layouts of `TBuf` and `TBufC`.

These descriptors are useful for fixed-size, relatively small strings, since they are stack-based. They may be considered equivalent to `char[]` in C, but with the benefit of overflow checking.

`TBufC` is the non-modifiable buffer class, used to hold constant string or binary data. The class derives from `TBufCBase` (which derives from `TDesC`, and exists as an inheritance convenience rather than to be used directly). `TBufC<n>` is a thin template class which uses an integer value to determine the size of the data area allocated for the buffer descriptor object.

`TBufC` defines several constructors that allow non-modifiable buffers to be constructed from a copy of any other descriptor, or from a zero-terminated string. They can also be created empty and filled later, since, although the data is non-modifiable, the entire contents of the buffer may be replaced by calling the assignment operator defined by the class. The replacement data may be another non-modifiable descriptor or a zero-terminated string, but in each case the new data length must not exceed the length specified in the template parameter when the buffer was created.

```
_LIT(KPalindrome, "Satan, oscillate my metallic sonatas");
TBufC<50> buf1(KPalindrome); // Constructed from literal descriptor
TBufC<50> buf2(buf1);         // Constructed from buf1
// Constructed from a NULL-terminated C string
```

```
TBufC<30> buf3((TText16*)"Never odd or even");
TBufC<50> buf4; // Constructed empty, length = 0
// Copy and replace
buf4 = buf1; // buf4 contains data copied from buf1, length modified
buf1 = buf3; // buf1 contains data copied from buf3, length modified
buf3 = buf2; // Panic! Max length of buf3 is insufficient for buf2 data
```

The `TBuf<n>` class for modifiable buffer data is a thin template class (see Section 3.4), the integer value determining the maximum allowed length of the buffer. It derives from `TBufBase`, which itself derives from `TDes`, thus inheriting the full range of descriptor operations in `TDes` and `TDesC`.

`TBuf<n>` defines a number of constructors and assignment operators, similar to those offered by its non-modifiable counterpart, `TBufC<n>`.

```
_LIT(KPalindrome, "Satan, oscillate my metallic sonatas");
TBuf<40> buf1(KPalindrome); // Constructed from literal descriptor
TBuf<40> buf2(buf1);        // Constructed from constant buffer descriptor
TBuf8<40> buf3((TText8*)"Do Geese see God?"); // from C string
TBuf<40> buf4; // Constructed empty, length = 0, maximum length = 40
// Copy and replace
buf4 = buf2; // buf2 copied into buf4, updating length and max length
buf3 = (TText8*)"Murder for a jar of red rum"; // updated from C string
```

Dynamic descriptors: `HBufC`

Heap-based descriptors can be used for string data that cannot be placed on the stack because it is too big, or because its size is not known at compile time. They are used where `malloc`'d data would be used in C.

The `HBufC8` and `HBufC16` classes (and the neutral version `HBufC`, which is `typedef`'d to `HBufC16`) export a number of static `NewL()` functions to create the descriptor on the heap. These methods follow the two-phase construction model (see Chapter 6) and may leave if there is insufficient memory available. There are no public constructors, and all heap buffers must be constructed using one of these methods (or from one of the `Alloc()` or `AllocL()` methods of the `TDesC` class, which spawn an `HBufC` copy of any existing descriptor).

As the C in the class name indicates, these descriptors are not modifiable, although, in common with the stack-based non-modifiable buffer descriptors, the class provides a set of assignment operators to allow the entire contents of the buffer to be replaced. Objects of the class can also be modified at run-time by creating a modifiable pointer descriptor, `TPtr`, using the `Des()` method (see Section 7.4).

```
_LIT(KPalindrome, "Do Geese see God?");
TBufC<20> stackBuf(KPalindrome);
// Allocates an empty heap descriptor of max length 20
HBufC* heapBuf = HBufC::NewLC(20);
```

```
TInt length = heapBuf->Length();// Current length = 0
TPtr ptr(heapBuf->Des());        // Modification of the heap descriptor
ptr = stackBuf; // Copies stackBuf contents into heapBuf
length = heapBuf->Length();      // length = 17
HBufC* heapBuf2 = stackBuf.AllocLC(); // From stack buffer
length = heapBuf2->Length();           // length = 17
_LIT(KPalindrome2, "Palindrome");
*heapBuf2 = KPalindrome2;      // Copy and replace data in heapBuf2
length = heapBuf2->Length(); // length = 10
CleanupStack::PopAndDestroy(2, heapBuf);
```

The heap descriptors can be created dynamically to the size required, but they are not automatically resized. The buffer must have sufficient memory available for the modification operation to succeed, or a panic will occur.

Dynamic descriptors: `RBuf`

Class `RBuf` behaves like `HBufC` in that the maximum length required can be specified dynamically. On instantiation, an `RBuf` object can allocate its own buffer or take ownership of pre-allocated memory or a pre-existing heap descriptor. `RBuf` descriptors are typically created on the stack, but maintain a pointer to memory on the heap.

`RBuf` is derived from `TDes`, so an `RBuf` object can easily be modified and can be passed to any function where a `TDesC` or `TDes` parameter is specified. This means that there is no need to create a `TPtr` around the data in order to modify it, which makes it preferable to `HBufC` when dynamically allocating a descriptor which is later modified.

Here are the public methods exported from class `RBuf16` (from `e32des16.h`) – the private and protected methods and member data of the class have been omitted for clarity.

```
class RBuf16 : public TDes16
  {
public:
  IMPORT_C RBuf16();
  IMPORT_C explicit RBuf16(HBufC16* aHBuf);
  IMPORT_C void Assign(const RBuf16& aRBuf);
  IMPORT_C void Assign(TUint16 *aHeapCell,TInt aMaxLength);
  IMPORT_C void Assign(TUint16 *aHeapCell,TInt aLength,TInt aMaxLength);
  IMPORT_C void Assign(HBufC16* aHBuf);
  IMPORT_C void Swap(RBuf16& aRBuf);
  IMPORT_C TInt Create(TInt aMaxLength);
  IMPORT_C void CreateL(TInt aMaxLength);
  IMPORT_C TInt CreateMax(TInt aMaxLength);
  IMPORT_C void CreateMaxL(TInt aMaxLength);
  inline void CreateL(RReadStream &aStream,TInt aMaxLength);
  IMPORT_C TInt Create(const TDesC16& aDes);
  IMPORT_C void CreateL(const TDesC16& aDes);
  IMPORT_C TInt Create(const TDesC16& aDes,TInt aMaxLength);
  IMPORT_C void CreateL(const TDesC16& aDes,TInt aMaxLength);
```

```
IMPORT_C TInt ReAlloc(TInt aMaxLength);
IMPORT_C void ReAllocL(TInt aMaxLength);
IMPORT_C void Close();
IMPORT_C void CleanupClosePushL();
};
```

Internally, RBuf behaves in one of two ways:

- As a TPtr descriptor type, which points to a buffer containing only descriptor data (the RBuf object allocates or takes ownership of memory existing elsewhere)
- As a pointer to a heap descriptor, HBufC* (the RBuf object takes ownership of an existing heap descriptor, thus the object pointed to contains a complete descriptor object).

Figure 7.3 shows both possible internal representations of an RBuf object initialized with a string of five characters representing the English word ''Hello''.

However, the handling of this distinction is transparent and there is no need to know how a specific RBuf object is represented internally, whether as a TPtr or as a pointer to HBufC. Calling the descriptor operations is straightforward because they correspond to the usual base-class methods of TDes and TDesC16. Section 7.6 describes the construction and usage of RBuf objects in more detail.

The class is not named HBuf because, unlike HBufC, objects of this type are not themselves directly created on the heap. It is instead an R class, because it manages a heap-based resource and is responsible for freeing the memory at cleanup time.

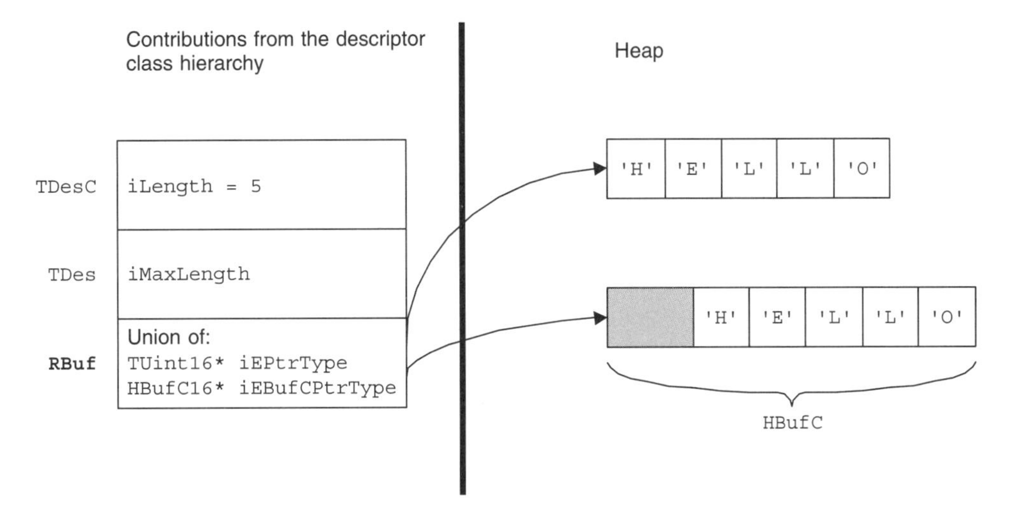

Figure 7.3 The two possible memory layouts for an RBuf descriptor

Exam Essentials

- Know the characteristics of the `TDesC`, `TDes`, `TBufC`, `TBuf`, `TPtrC`, `TPtr`, `RBuf` and `HBufC` descriptor classes
- Understand that the descriptor base classes `TDesC` and `TDes` implement all generic descriptor manipulation code, while the derived descriptor classes merely add construction and assignment code specific to their type
- Identify the correct and incorrect use of modifier methods in the `TDesC` and `TDes` classes
- Recognize that there is no `HBuf` class, but that `RBuf` can be used instead as a modifiable dynamically allocated descriptor

7.3 The Inheritance Hierarchy of the Descriptor Classes

Figure 7.4 shows the inheritance hierarchy of the descriptor classes. The `TDesC` and `TDes` base classes provide the descriptor manipulation methods, and must know the type of derived class they are operating on in order to correctly locate the data area. However, each subclass does not implement its own data access method using virtual function overriding, because this would add an extra 4 bytes to each derived descriptor object for a virtual pointer (*vptr*) to access the virtual function table. Descriptors were designed to be as efficient as possible, and the size overhead to accommodate a C++ vptr was considered undesirable.

Instead, to allow for the specialization of derived classes, the top 4 bits of the 4 bytes that store the length of the descriptor object are used to indicate the class type of the descriptor. There are currently six derived descriptor classes (`TPtrC`, `TPtr`, `TBufC`, `TBuf`, `HBufC` and `RBuf`), each of which sets the identifying bits as appropriate upon construction. The use of 4 bits to identify the type limits the number of different types of descriptor to 2^4 (= 16), but since only six types have been necessary in current and previous releases of Symbian OS, it seems unlikely that the range will need to be extended significantly in the future.

Access to the descriptor data for all descriptors goes through the non-virtual `Ptr()` method of the base class `TDesC`, which uses a `switch` statement to check the 4 bits, identify the type of descriptor and return the correct address for the beginning of its data. This requires that the `TDesC` base class has knowledge of the memory layout of its subclasses hard-coded into `Ptr()`. Section 7.2 describes the different memory layouts for pointer and buffer descriptors in more detail.

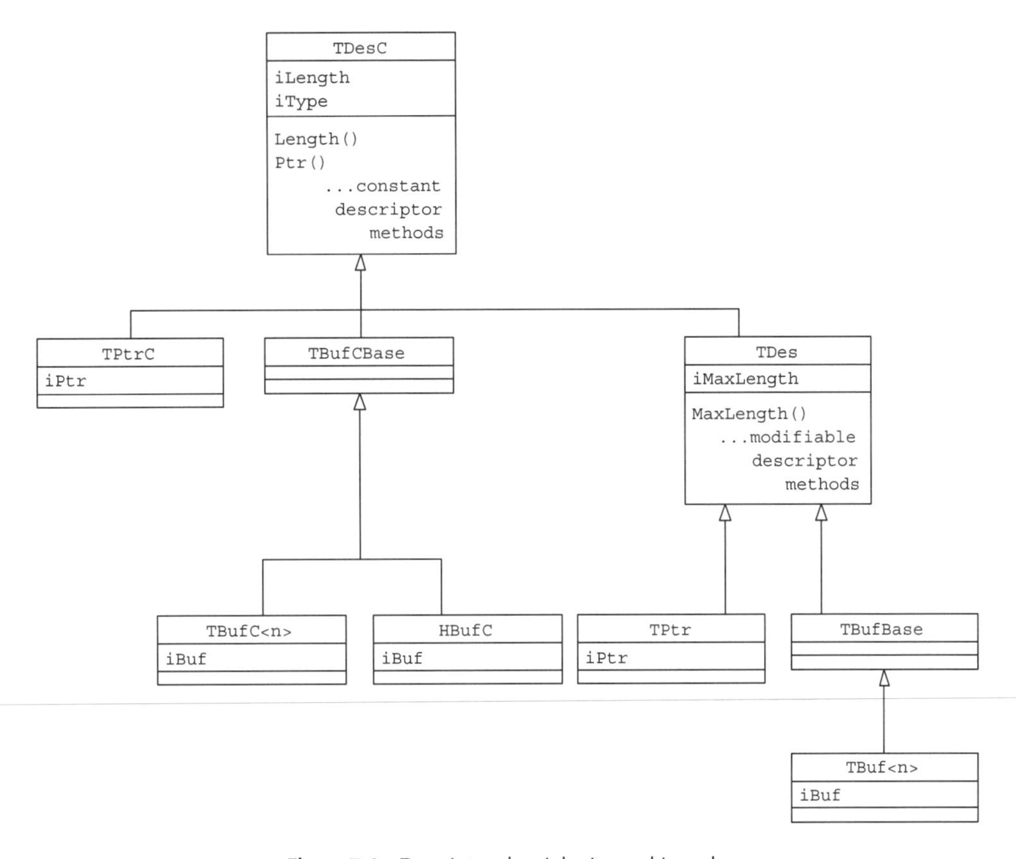

Figure 7.4 Descriptor class inheritance hierarchy

Exam Essentials

- Know the inheritance hierarchy of the descriptor classes
- Understand the memory efficiency of the descriptor class inheritance model and its implications

7.4 Using the Descriptor APIs

As described in Section 7.2, the `TDesC` and `TDes` descriptor base classes provide and implement the APIs for all descriptor operations, regardless of the actual type of the derived descriptor. Typically, the derived descriptors only implement specific methods for construction and copy assignment.

However, objects of type `TDesC` and `TDes` cannot be instantiated directly because their default constructors are protected. It is the derived descriptor types that are actually instantiated and used. The descriptor API methods are fully documented in the Symbian OS Library of each SDK. The following discussion concentrates on some of the trickier areas of descriptor manipulation.

The difference between `Length()` *and* `Size()`

The `Size()` method returns the size of the descriptor in bytes. The `Length()` method returns the number of characters it contains. For 8-bit descriptors, this is the same as the size, because the size of a character is a byte. However, from Symbian OS v5u the native character is 16 bits wide, which means that each character occupies two bytes. For this reason, `Size()` always returns a value double that of `Length()` for neutral and explicitly wide descriptors.

`MaxLength()` *and length modification methods*

The `MaxLength()` method of `TDes` returns the maximum length of a modifiable descriptor value. Like the `Length()` method of `TDesC`, it is not overridden by the derived classes. The `SetMax()` method doesn't (as one might think) change the maximum length of the descriptor, thereby expanding or contracting the data area; instead, it sets the current length of the descriptor to the maximum length allowed.

The `SetLength()` method can be used to adjust the descriptor length to any value between zero and its maximum length, inclusively. The `Zero()` method sets the length to zero.

`Set()` *and the assignment operator*

The pointer descriptor types provide `Set()` methods, which update them to point at different string data.

```
// Literal descriptors are described later in this chapter
_LIT(KDes1, "Sixty zippers were quickly picked from the woven jute bag");
_LIT(KDes2, "Waltz, bad nymph, for quick jigs vex");
TPtrC alpha(KDes1);
TPtrC beta(KDes2);
alpha.Set(KDes2); // alpha points to the data in KDes2
beta.Set(KDes1);  // beta points to the data in KDes1
```

Additionally, `TDes` provides an assignment operator, =, to copy data into the memory referenced by any modifiable descriptor (as long as the length of the data to be copied does not exceed the maximum length of the descriptor, which would cause a panic).

It is easy to confuse `Set()`, which resets a pointer descriptor to point at a new data area (with corresponding modification to the length and maximum length members), with `TDes::operator=()`, which merely copies data into an existing descriptor (and may modify the descriptor length but not its maximum length).

```
_LIT(KLiteralDes1, "Jackdaws love my big sphinx of quartz");
TBufC<60> buf(KLiteralDes1);
TPtr ptr(buf.Des());       // Points to the contents of buf
TUint16* memoryLocation; // Valid pointer into memory
...
TInt maxLen = 40; // Maximum length to be represented
TPtr memPtr(memoryLocation, maxLen); // max length=40
// Copy and replace
memPtr = ptr; // memPtr data is KLiteralDes1 (37 bytes), maxLength=40
_LIT(KLiteralDes2, "The quick brown fox jumps over the lazy dog");
TBufC<100> buf2(KLiteralDes2);
TPtr ptr2(buf2.Des()); // Points to the data in buf
// Replace what ptr points to
ptr.Set(ptr2); // ptr points to contents of buf2, max length = 100
memPtr = ptr2; // Attempt to update memPtr, which panics because the
// contents of ptr2 (43 bytes) exceeds max length of memPtr (40 bytes)
```

Des()

The stack- and heap-based constant buffer descriptor classes, `TBufC` and `HBufC`, provide a method which returns a modifiable pointer descriptor to the data represented by the buffer. So, while the content of a non-modifiable buffer descriptor cannot be altered directly, other than by complete replacement of the data, it is possible to change the data indirectly by calling `Des()` and then operating on the data via the pointer. When the data is modified via the return value of `Des()`, the length members of both the pointer descriptor and the constant buffer descriptor are updated.

For `TBufC`:

```
_LIT8(KPalindrome, "Satan, oscillate my metallic sonatas");
TBufC8<40> buf(KPalindrome); // Constructed from literal descriptor
TPtr8 ptr(buf.Des()); // data is the string in buf, max length = 40
// Use ptr to replace contents of buf
ptr = (TText8*)"Do Geese see God?";
ASSERT(ptr.Length()==buf.Length());
_LIT8(KPal2, "Are we not drawn onward, we few, drawn onward to new era?");
ptr = KPal2; // Panic! KPal2 exceeds max length of buf (=40)
```

The use of `Des()` to modify the contents of `HBufC` was illustrated in Section 7.2.

Exam Essentials

- Understand that the descriptor base classes `TDesC` and `TDes` cannot be instantiated

- Understand the difference between `Size()`, `Length()` and `MaxLength()` descriptor methods
- Understand the difference between `Copy()` and `Set()` descriptor methods and how to use assignment correctly

7.5 Descriptors as Function Parameters

Since the `TDesC` and `TDes` descriptor base classes provide and implement the APIs for all descriptor operations, they can be used as function arguments and return types, allowing descriptors to be passed around in code without forcing a dependency on a particular type. An API client shouldn't be constrained to using a `TBuf` just because a particular function requires it, and function providers will remain agnostic to the type of descriptor passed to them. Unless a function takes or returns ownership, it shouldn't even need to specify whether a descriptor is stack- or heap-based.

The descriptor APIs allow for this independence as long as the descriptor is one of the standard types, so that the appropriate descriptor methods can be called on it. For this reason, when defining functions, the abstract base classes should always be used as parameters or return values. For efficiency, descriptor parameters should be passed by reference, either as const `TDesC&` for constant descriptors or `TDes&` when modifiable.

Exceptions

When returning ownership of a heap-based descriptor as a return value, it should be specified explicitly so that the caller can clean it up appropriately and avoid a memory leak

Exam Essentials

- Understand that the correct way to specify a descriptor as a function parameter is to use a reference, for both constant data and data that may be modified by the function in question.

7.6 Correct Use of the Dynamic Descriptor Classes

`HBufC` construction and usage

`HBufC` can be spawned from existing descriptors using the `Alloc()` or `AllocL()` overloads implemented by `TDesC`. Here is a contrived example which shows how to replace inefficient code with use of `TDesC::AllocL()`:

```
void CSampleClass::UnnecessaryCodeL(const TDesC& aDes)
  {
  iHeapBuffer = HBufC::NewL(aDes.Length());
  TPtr ptr(iHeapBuffer->Des());
  ptr.Copy(aDes);
  ...
  // could be replaced by a single line
  iHeapBuffer = aDes.AllocL();
  }
```

An `HBufC` object can also be instantiated using the static `NewL()` factory methods specified for the class. For `HBufC16`, the methods available are as follows:

```
static IMPORT_C HBufC16* NewL(TInt aMaxLength);
static IMPORT_C HBufC16* NewLC(TInt aMaxLength);
```

These methods create a new heap-based buffer descriptor with maximum length as specified, leaving if there is insufficient memory available for the allocation. The latter method leaves the successfully created descriptor object on the cleanup stack (see Chapter 5). The heap descriptor is empty and its length is set to zero.

```
static IMPORT_C HBufC16* New(TInt aMaxLength);
```

This method creates a new heap-based buffer descriptor with maximum length as specified. It does not leave if there is no heap memory available to allocate the descriptor, so the caller must compare the returned pointer against `NULL` to confirm that it has succeeded before dereferencing it. The heap descriptor is empty and its length is set to zero.

```
static IMPORT_C HBufC16* NewMax(TInt aMaxLength);
static IMPORT_C HBufC16* NewMaxL(TInt aMaxLength);
static IMPORT_C HBufC16* NewMaxLC(TInt aMaxLength);
```

These three methods create a new heap-based buffer descriptor with maximum length as specified and set its length to the maximum value, although no data is assigned to the descriptor. The methods either return a `NULL` pointer (`NewMax()`) or leave (`NewMaxL()`, `NewMaxLC()`) if insufficient memory is available for the allocation. `NewMaxLC()` leaves the successfully allocated descriptor on the cleanup stack.

```
static IMPORT_C HBufC16* NewL(RReadStream& aStream, TInt aMaxLength);
static IMPORT_C HBufC16* NewLC(RReadStream& aStream, TInt aMaxLength);
```

These methods allocate a heap-based buffer descriptor and initialize it from the contents of a read stream, by reading from the stream (up to the maximum length specified) and allocating a buffer to hold the contents. They are typically used to reconstruct a descriptor that has previously been externalized to a write stream using the stream operators.

The following code shows a naive implementation of descriptor externalization and internalization, using four bytes to store the descriptor length, then adding the descriptor data to the stream separately. In the `InternalizeL()` method, the descriptor is reconstructed in four rather awkward stages.

```
// Writes the contents of iHeapBuffer to a writable stream
void CSampleClass::ExternalizeL(RWriteStream& aStream) const
  {
  // Write the descriptor's length
  aStream.WriteUint32L(iHeapBuffer->Length());
  // Write the descriptor's data
  aStream.WriteL(*iHeapBuffer, iHeapBuffer->Length());
  }
// Instantiates iHeapBuffer by reading the contents of the stream
void CSomeClass::InternalizeL(RReadStream& aStream)
  {
  TInt size=aStream.ReadUint32L(); // Read the descriptor's length
  iHeapBuffer = HBufC::NewL(size); // Allocate iHeapBuffer

  // Create a modifiable descriptor over iHeapBuffer
  TPtr ptr(iHeapBuffer->Des());
  // Read the descriptor data into iHeapBuffer
  aStream.ReadL(ptr,size);
  }
```

The following code shows the use of the stream operators as a more efficient alternative. The stream operators have been optimized to compress the descriptor metadata as much as possible for efficiency and space conservation (see Chapter 12).

```
void CSampleClass::ExternalizeL(RWriteStream& aStream) const
  {// Much more efficient, no wasted storage space
  aStream << iHeapBuffer;
  }
void CSampleClass::InternalizeL(RReadStream& aStream)
  {// KMaxLength indicates the maximum length of
   // data to be read from the stream
  iHeapBuffer = HBufC::NewL(aStream, KMaxLength);
  }
```

`RBuf` construction and usage

While there is a non-modifiable heap descriptor class, `HBufC`, there is no corresponding modifiable `HBuf` class, which might have been expected in order to make heap buffers symmetrical with `TBuf` stack buffers. However, since Symbian OS v8.1, the `RBuf` class has been available for use as a modifiable heap-based buffer class.

RBuf objects can be instantiated using Create(), CreateMax() or CreateL() to specify the maximum length of descriptor data that can be stored. It's also possible to instantiate an RBuf and copy the contents of another descriptor into it, as follows:

```
RBuf myRBuf;
_LIT(KHelloRBuf, "Hello RBuf!");
myRBuf.Create(KHelloRBuf());
```

Create() allocates a buffer for the RBuf to reference. If that RBuf previously owned a buffer, Create() will not clean it up before assigning the new buffer reference, so this must be done explicitly by calling Close() first to free any pre-existing owned allocated memory.

Alternatively, an RBuf can be instantiated to take ownership of a pre-existing section of memory using the Assign() method.

```
// Taking ownership of HBufC
HBufC* myHBufC = HBufC::NewL(20);
RBuf myRBuf.Assign(myHBufC);
... // Use and clean up

// Taking ownership of pre-allocated heap memory
TInt maxSizeOfData = 20;
RBuf myRBuf;
TUint16* pointer = static_cast<TUint16*>(User::AllocL(maxSizeOfData*2));
myRBuf.Assign(pointer, maxSizeOfData);
... // Use and clean up
```

As for the Create() methods described above, Assign() will also orphan any data already owned by the RBuf. Close() should be called where appropriate to avoid memory leaks.

The RBuf class doesn't manage the size of the buffer and reallocate it if more memory is required for a particular operation. If Append() is called on an RBuf object for which there is insufficient memory available, a panic will occur. This should be clear from the fact that the base-class methods are non-leaving, meaning there is no scope for the reallocation to fail in the event of low memory.

As with the other descriptor classes, the memory for descriptor operations must be managed by the programmer. The ReAllocL() method can be used as follows:

```
// myRBuf is the buffer to be resized e.g. for an Append() operation
myRBuf.CleanupClosePushL(); // push onto cleanup stack for leave-safety
myRBuf.ReAllocL(newLength); // extend to newLength
CleanupStack::Pop ();       // remove from cleanup stack
```

Using an RBuf is preferable to using an HBufC in that, if the ReAllocL() method is used on the HBufC and causes the heap cell to move, any associated `HBufC*` and `TPtr` variables need to be updated. This update isn't required for `RBuf` objects, since the pointer is maintained internally.

The class is not named `HBuf` because, unlike `HBufC`, objects of this type are not themselves directly created on the heap. It is instead an R class, because it manages a heap-based resource and is responsible for freeing the memory at cleanup time. As is usual for other R classes, cleanup is performed by calling `Close()` (or `CleanupStack::PopAndDestroy()` if the `RBuf` was pushed onto the cleanup stack by a call to `CleanupClosePushL()`). See Chapter 4 for a full discussion of Symbian OS R-class types.

It's possible to create an `RBuf` from an existing `HBufC`, making it easy to migrate code to use the new class. Because the `RBuf` class is both modifiable and dynamically allocated, it is desirable to use it where previously it would have been necessary to instantiate an `HBufC` and then use a companion `TPtr` object, constructed by calling `Des()` on the heap descriptor. As can be seen from the following example, the resulting code is simpler and shorter.

Code which previously was written as follows:

```
HBufC* socketName = NULL;
...
if(!socketName)
  {
  socketName = HBufC::NewL(KMaxNameLength); // KMaxNameLength is defined
                                            // elsewhere
  }
TPtr socketNamePtr(socketName->Des()); // Create writable 'companion' TPtr
message.ReadL(message.Ptr0(), socketNamePtr);
```

can be converted to the following:

```
RBuf socketName;
...
if(socketName.Compare(KNullDesC)==0)
  {
  socketName.CreateL(KMaxNameLength);
  }
message.ReadL(message.Ptr0(), socketName);
```

Because the code is simpler, it is easier to understand and maintain. For this reason, `RBuf` is recommended for use when a dynamically allocated buffer is required to hold data that changes frequently. `HBufC` should be preferred when a dynamically allocated descriptor is needed to hold data that rarely changes.

The `RBuf` class was first introduced in Symbian OS v8.0, but first documented in Symbian OS v8.1 and used most extensively in software designed for phones based on Symbian OS v9 and later.

Exam Essentials

- Identify the correct techniques and methods to instantiate an `HBufC` heap buffer object
- Recognize and demonstrate knowledge of how to use the new descriptor class `RBuf`

7.7 Common Inefficiencies in Descriptor Usage

`TFileName` objects waste stack space

The `TFileName` type is `typedef`'d as a modifiable stack buffer with maximum length 256 characters. It can be useful when calling various file system functions to parse filenames into complete paths, for example to print out a directory's contents, since the exact length of a filename isn't always known at compile time.

However, since each character is of 16-bit width, every time a `TFileName` object is declared on the stack, it consumes $2 \times 256 = 512$ bytes (plus the 12 bytes required for the descriptor object itself). That's just over 0.5 KB.

On Symbian OS, the stack space for each process is limited; by default it is just 8 KB. On the Windows emulator, if more stack space is needed than the default, the stack will just expand. This is not the case on a phone, however – if more stack is used than is available, a panic will be raised when the stack overflow occurs. This can be hard to track down, since it will not be seen when testing on the emulator, so cannot be easily debugged.

It is clear that a single `TFileName` object can consume, and potentially waste, a significant proportion of the stack space. It's good practice to use one of the dynamic heap descriptor types instead, or limit the use of `TFileName` objects to members of C classes, since these types are always created on the heap.

Referencing `HBufC` through `TDesC`

The `HBufC` class derives from `TDesC` and an `HBufC*` pointer can simply be dereferenced when a reference to a non-modifiable descriptor (`TDesC&`) is required. A common mistake is to call the `Des()` method on the heap descriptor, creating a separate `TPtr` referencing the descriptor data. This is not incorrect (it returns a `TDes&`), but it is clearer and more efficient simply to return the `HBufC` object directly.

```
const TDesC& CSampleClass::AccidentalComplexity()
  {
  return (iHeapBuffer->Des());
  // could be replaced more efficiently with
```

```
return (*iHeapBuffer);
}
```

Exam Essentials

- Know that `TFileName` objects should not be used indiscriminately, because of the stack space each consumes
- Understand when to dereference an `HBufC` object directly, and when to call `Des()` to obtain a modifiable descriptor (`TDes&`)

7.8 Literal Descriptors

Literal descriptors are somewhat different from the other descriptor types. They are equivalent to `static char[]` in C and can be built into program binaries in ROM because they are constant. There is a set of macros defined in `e32def.h` which can be used to define Symbian OS literals of two different types, `_LIT` and `_L`.

_LIT macro

The `_LIT` macro is preferred for Symbian OS literals, since it is the more efficient type. It has been used in the sample code throughout this chapter, typically as follows:

```
_LIT(KFieldMarshalTait, "Field Marshal Tait");
```

`KFieldMarshalTait` can then be used as a constant descriptor, for example to write to a file or display to a user. The `_LIT` macro builds a named object (`KFieldMarshalTait`) of type `TLitC16` into the program binary, storing the appropriate string (in this case, `"Field Marshal Tait"`). The explicit macros `_LIT8` and `_LIT16` behave similarly except that `_LIT8` builds a narrow string of type `TLitC8`.

`TLitC8` and `TLitC16` do not derive from `TDesC8` or `TDesC16`, but they have the same binary layouts as `TBufC8` or `TBufC16`. This allows objects of these types to be used wherever `TDesC` is used. In fact, the string stored in the program binary has a `NULL` terminator because the native compiler string is used to build it. However, the `_LIT` macro adjusts the length to the correct value for a non-terminated descriptor.

Symbian OS also defines literals to represent a blank string. There are three variants of the *null descriptor*, defined as follows:

```
// Build independent:
_LIT(KNULLDesC,"");
```

```
// 8-bit for non-Unicode strings:
_LIT8(KNULLDesC8,"");
// 16-bit for Unicode strings:
_LIT16(KNULLDesC16,"");
```

***_L* macro**

Use of the _L macro is now deprecated in production code, though it may still be used in test code (where memory use is less critical). The advantage of using _L (or the explicit forms _L8 and _L16) is that it can be used in place of a TPtrC without having to declare it separately from where it is used:

```
User::Panic(_L("telephony.dll"), KErrNotSupported);
```

For the example above, the string ("telephony.dll") is built into the program binary as a basic, NULL-terminated string, with no initial length member, unlike the TLitC built for the _LIT macro. Because there is no length word, the layout of the stored literal is not like that of a descriptor and, when the code executes, each instance of _L will result in construction of a temporary TPtrC, with the pointer set to the address of the first byte of the literal as it is stored in ROM. The use of such a run-time temporary is safe as long as it is used only during the lifetime of the function in which it is created, or if it is copied for use outside of that lifetime. However, the construction of a temporary, which requires setting the pointer, the length and the descriptor type, is an overhead in terms of inline constructor code which may bloat binaries where many string literals are used. This is why the _LIT macro is considered preferable.

Figure 7.5 shows the difference in memory layout in ROM for literals created using _LIT and _L macros.

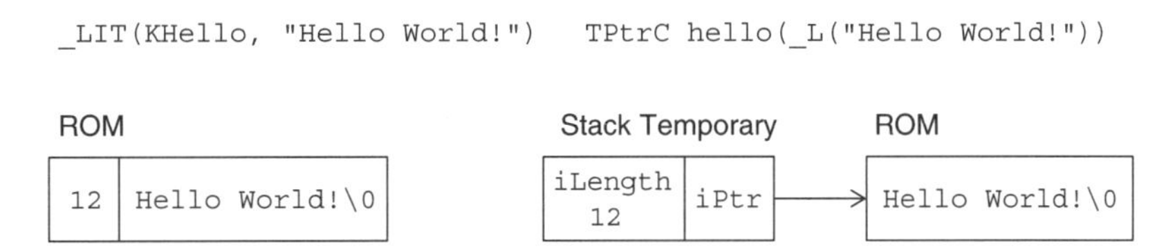

Figure 7.5 Memory layout for literal descriptors

Exam Essentials

- Know how to manipulate literal descriptors and know that those specified using _L are deprecated
- Specify the difference between literal descriptors using _L and those using _LIT and the disadvantages of using the former

7.9 Descriptor Conversion

Conversion between narrow and wide descriptors

TDes implements an overloaded set of Copy() methods which allow copying directly into descriptor data from another descriptor, from a NULL-terminated string or from a pointer. The Copy() methods copy the data into the descriptor, setting its length accordingly; the methods will panic if the maximum length of the receiving descriptor is shorter than the incoming data.

The Copy() method is overloaded to take either an 8- or 16-bit descriptor. Thus, not only is it possible to copy a narrow-width descriptor onto a narrow-width descriptor and a wide descriptor onto a wide descriptor, but it is also possible to copy between descriptor widths, effecting a conversion of sorts in the process.

As Figure 7.6 illustrates, the Copy() method implemented by TDes8 to copy an incoming wide descriptor into a narrow descriptor strips out alternate characters, assuming them to be zeroes, that is, that the data values do not exceed 255 (decimal). The Copy() method which copies a narrow descriptor into the data area is a straight data copy.

```
// Instantiate a narrow descriptor
TBuf8<3> cat(_L8("CAT")); // _L8 is described in Section 7.8
// Instantiate a wide descriptor
TBuf16<3> dog(_L16("DOG"));
// Copy the contents of the wide descriptor into the narrow descriptor
cat.Copy(dog); // cat now contains "DOG"
```

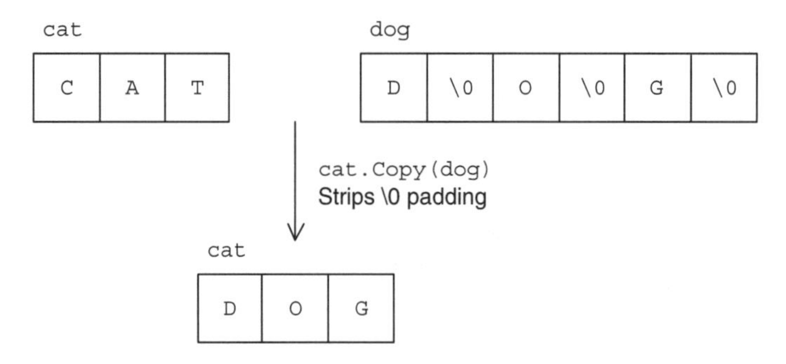

Figure 7.6 NULL characters stripped out in wide-to-narrow descriptor copy

Conversely, for class TDes16, an incoming 16-bit descriptor can be copied directly onto the data area, but the Copy() method that takes an 8-bit descriptor pads each character with a trailing zero as part of the copy operation, as shown in Figure 7.7.

```
// Instantiate a narrow descriptor
TBuf8<5> small(_L8("SMALL"));
// Instantiate a wide descriptor
TBuf16<5> large(_L16("LARGE"));
// Copy the contents of the narrow descriptor into the wide descriptor
large.Copy(small); // large now contains "SMALL"
```

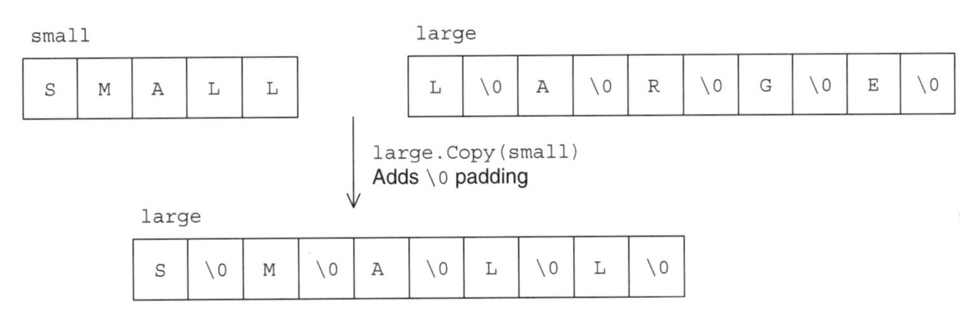

Figure 7.7 NULL characters used to pad narrow-to-wide descriptor copy

The `Copy()` methods thus form a rudimentary means of copying and converting when the character set is encoded by one byte per character and the last byte of each wide character is simply a `NULL` character padding. To perform proper conversion in both directions between 16-bit Unicode (UCS-2) and 8-bit, non-Unicode, character sets (or between Unicode and the UTF-7 and UTF-8 transformation sets) the `CnvUtfConverter` class supplied by `charconv.lib` (see header file `utf.h`) is available. The class supplies a set of static methods such as `ConvertFromUnicodeToUtf8()` and `ConvertToUnicodeFromUtf8()`.

Converting descriptors to numbers

A descriptor can be converted to a number using the `TLex` class. This class provides general-purpose lexical analysis and performs syntactical element parsing and string-to-number conversion, using the locale-dependent functions of `TChar` to determine whether each character is a digit, a letter or a symbol.

Like the descriptor classes, `TLex` is the build-width neutral class (implicitly `TLex16`) while `TLex8` and `TLex16` can also be used explicitly. The neutral form should be preferred unless a particular variant is required.

`TLex` can be constructed with the data for lexical analysis or constructed empty and later assigned the data. Both construction and assignment can take another `TLex` object, a non-modifiable descriptor, or a `TUint16*` or `TUint8*` (for `TLex16` or `TLex8` respectively) pointer to string data.

At the very simplest level, when the string contains just numerical data, the descriptor contents can be converted to an integer using the `Val()` function of `TLex`.

```
_LIT(KTestLex, "54321");
TLex lex(KTestLex());
TInt value = 0;
TInt err = lex.Val(value)); // value == 54321 if no error occurred
```

The `Val()` function is overloaded for different signed integer types (`TInt`, `TInt8`, `TInt16`, `TInt32`, `TInt64`), with or without limit checking. There are also `Val()` overloads for the unsigned integer types, passing in a radix value (decimal, hexadecimal, binary or octal), and for `TReal`.

`TLex` provides a number of other API methods for manipulation and parsing, which are documented in the Symbian Developer Library.

Converting numbers to descriptors

The descriptor classes provide several ways to convert a number to a descriptor. The various overloads of `AppendNum()`, `AppendNumFixedWidth()`, `Num()` and `NumFixedWidth()` convert the specified numerical parameter to a character representation and either replace the contents of, or append the data to, the descriptor on which the method is called.

The `Format()`, `AppendFormat()`, `FormatList()` and `AppendFormatList()` methods of `TDes` each take a format string, containing literal text embedded with conversion directives, and a trailing list of arguments. Each formatting directive consumes one or more arguments from the trailing list and can be used to convert numbers into descriptor data.

Packaging objects in descriptors

Flat data objects can be stored conveniently within descriptors using the package buffer (`TPckgBuf`) and package pointer (`TPckg` and `TPckgC`) classes. This is useful for inter-thread or inter-process data transfer when making a client–server request (see Chapter 11). In effect, a T-class object may be packaged whole into a descriptor ("descriptorized") so it may be passed easily in a type-safe way between threads or processes.

The `TPckgBuf`, `TPckg` and `TPckgC` classes are thin template classes (see Chapter 3) derived from `TBuf<n>`, `TPtr` and `TPtrC` respectively (see `e32std.h`). The classes are type-safe and are templated on the type to be packaged.

There are two package pointer classes, creating either modifiable (`TPckg`) or non-modifiable (`TPckgC`) pointer descriptors which refer to the existing instance of the template-packaged class. Functions may be called on the enclosed object; if it is enclosed in a `TPckgC`, a constant reference to the packaged object is returned from operator `()`.

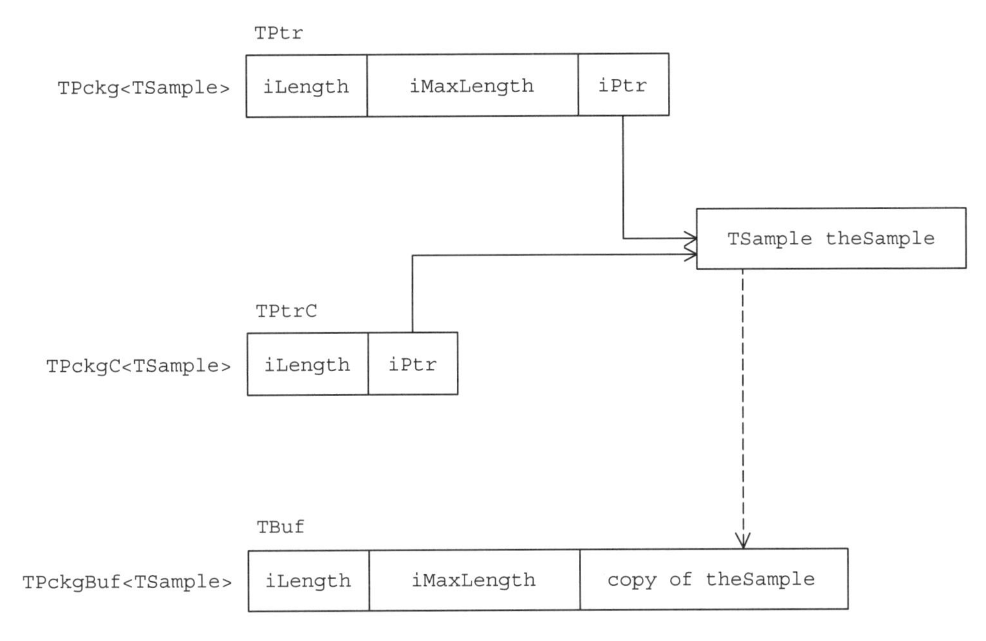

Figure 7.8 Memory layout of the `TPckg`, `TPckgC` and `TPckgBuf` classes

The package buffer `TPckgBuf` creates and stores a new instance of the type to be encapsulated in a descriptor. The copied object is owned by the package buffer; it is modifiable and functions may be called on it, after calling operator `()` on the `TPckgBuf` object to retrieve it. Because the package buffer contains a copy of the original object, if a modification function is called it is the copy that is modified – the original is unchanged.

The following code shows an object of a simple T class encapsulated in each of the package types, and Figure 7.8 illustrates the memory layout of each.

```
class TSample
  {
public:
  void SampleFunction();
  void ConstantSampleFunction() const;
private:
  TInt iSampleData;
  };
TSample theSample;
TPckg<TSample> packagePtr(theSample);
TPckgC<TSample> packagePtrC(theSample);
TPckgBuf<TSample> packageBuf(theSample);
packagePtr().SampleFunction();
packagePtrC().SampleFunction();// Compile error! Non-const function
packagePtrC().ConstantSampleFunction();
packageBuf().SampleFunction();
```

Exam Essentials

- Know how to convert 8-bit descriptors into 16-bit descriptors and vice versa using the descriptor `Copy()` method or the `CnvUtfConverter` class
- Recognize how to read data from file into an 8-bit descriptor and then 'translate' the data to 16-bit without padding, and vice versa
- Know how to use the `TLex` class to convert a descriptor to a number, and `TDes::Num()` to convert a number to a descriptor

References

[Babin 2005 Chapter 6]
[Harrison 2004 Chapter 1]
Shackman, Mark, "Introducing the `RBuf` Descriptor", v1.0, ***www.symbian.com/developer/techlib/papers/RBuf/introduction_to_RBuf_v1.0.pdf***
[Stichbury 2004 Chapters 5 and 6]
Symbian Developer Library, "Resizable buffer descriptors", ***www.symbian.com/developer/techlib/v9.1docs/doc_source/guide/Base-subsystem-guide/N10086/BuffersAndStrings/Descriptors/DescriptorsGuide2/ResizableBufferDescriptors.guide.html***

8

Dynamic Arrays

Introduction

Dynamic container classes are very useful for manipulating collections of data without needing to know in advance how much memory to allocate for their storage. They expand as elements are added to them and do not need to be created with a fixed size. This chapter discusses the two families of dynamic array classes provided by Symbian OS, as well as the fixed-length array class provided to wrap the standard C++ `[]` array.

8.1 Dynamic Arrays in Symbian OS

Critical Information

Conceptually, the logical layout of an array is linear, like a vector. However, the implementation of a dynamic array can either use a single heap cell as a "flat" buffer to hold the array elements, or allocate the array buffer in a number of segments, using a doubly-linked list to manage the segmented heap memory.

Contiguous flat buffers are typically used when high-speed pointer lookup is an important consideration and when array resizing is expected to be infrequent.

Segmented buffers are preferable for large arrays which are expected to resize frequently, or where elements are frequently inserted into or deleted from the array.

- Repeated reallocations of a single flat buffer may result in heap thrashing and copying.
- Insertion and deletion are typically more efficient with a segmented buffer than with a flat buffer, since it does not require that all the elements after the modification point be shuffled into a new position.

Symbian OS provides two distinct class families for creating and accessing dynamic arrays. The original array classes, from the early days of Symbian OS, are C classes. There are a number of different types of array class, all of which have names prefixed by "CArray", such as CArrayFixFlat, CArrayFixSeg and CArrayVarSeg. They are referred to generically as the "CArrayX" classes. The RArray and RPointerArray classes were introduced at a later stage.

CArrayX classes

The number of CArrayX classes makes this array family very flexible, although they have a significant performance overhead (see Section 8.2).

The naming scheme works as follows. For each class, the CArray prefix is followed by:

- Fix for elements which have the same length and are copied so they may be contained in the array buffer
- Var where the elements are of different lengths; each element is contained within its own heap cell and the array buffer contains pointers to the elements
- Pak for a packed array where the elements are of variable length; elements are copied into the array buffer, each element preceded by its length information
- Ptr for an array of pointers to CBase-derived objects.

Following this, the array class name ends with Flat (for example CArrayFixFlat), for classes which use an underlying flat buffer for the dynamic memory of the array, or Seg (for example CArrayPtrSeg), for those that use a segmented buffer.

Figure 8.1 illustrates the various memory layouts available.

The inheritance hierarchy of the CArrayX classes is fairly straightforward. All of the classes are C classes and thus ultimately derive from CBase. Each class is a thin template specialization (see Section 3.4) of one of the array base classes, CArrayVarBase, CArrayPakBase or CArrayFixBase. Thus, for example, CArrayVarSeg<class T> and CArrayVarFlat<class T> derive from CArrayVar<class T>, which is a template specialization of CArrayVarBase.

CArrayVarBase owns an object which derives from CBufBase, the dynamic buffer base class, and which is used to store the elements of the array. The object is a concrete instance of CBufFlat (a flat dynamic storage buffer) or CBufSeg (a segmented dynamic buffer).

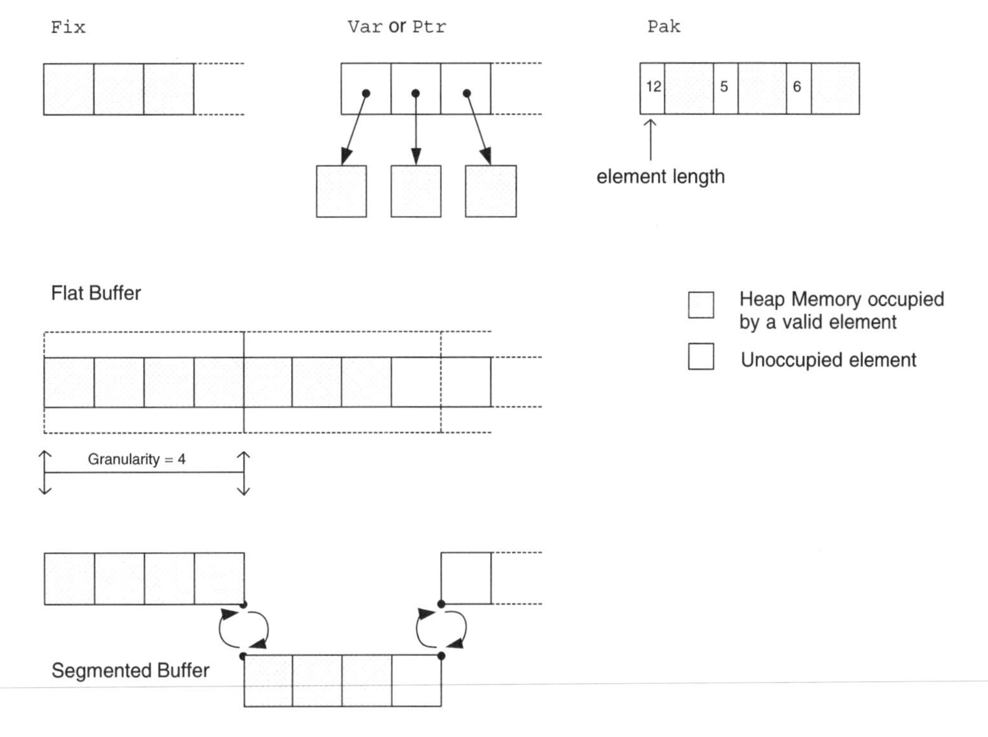

Figure 8.1 Memory layout of Symbian OS dynamic arrays

The following table summarizes the `CArrayX` classes available.

Array Class	Description	Cleanup behavior
`CArrayFixFlat`	Elements are of fixed size and are contained in the array itself. The array occupies a single area in memory.	Elements are owned and destroyed by the array.
`CArrayFixSeg`	Elements are of fixed size and are contained in the array itself. The array occupies multiple areas (segments) of memory.	Elements are owned and destroyed by the array.
`CArrayVarFlat`	Elements are of variable size. Each element exists	Elements are owned and destroyed by the array.

Array Class	Description	Cleanup behavior
	separately on the heap, and the array consists of pointers to those items. The array occupies a single area in memory.	Elements are owned and destroyed by the array.
`CArrayVarSeg`	Elements are of variable size. Each element exists separately on the heap, and the array consists of pointers to those items. The array occupies multiple areas (segments) of memory.	Elements are owned and destroyed by the array.
`CArrayPtrFlat`	Elements are pointers to `CBase`-derived objects. The array occupies a single area in memory.	Elements must be destroyed separately before array deletion by calling `ResetAndDestroy()`
`CArrayPtrSeg`	Elements are pointers to `CBase`-derived objects. The array occupies multiple areas (segments) of memory	Elements must be destroyed separately before array deletion by calling `ResetAndDestroy()`
`CArrayPakFlat`	The elements are of variable size but are contained within the array. The length of each element precedes the data, rather like a descriptor. The array occupies a single area in memory.	Elements are owned and destroyed by the array.

`RArray` and `RPointerArray`

`RArray` and `RPointerArray` are R classes (see Chapter 4), indicating that they own a resource, which in this case is the heap memory allocated to hold the array.

RArray objects themselves may be either stack- or heap-based. As with all R classes, the Close() or Reset() functions must be called to clean up properly, that is, to free the memory allocated for the array. RArray::Close() frees the memory used to store the array and closes it, while RArray::Reset() frees the memory associated with the array and resets its internal state, allowing the array to be reused. It is acceptable to call Reset() before allowing the array object to go out of scope, since all the heap memory associated with the object will have been cleaned up.

RArray<class T> is a thin template specialization of class RArrayBase and comprises a simple array of elements of the same size, which are stored within the array. To hold the elements the array class uses a flat, vector-like block of heap memory, which is resized when necessary.

RPointerArray<class T> is a thin template class deriving from RPointerArrayBase. It comprises a simple array of pointer elements and uses flat, linear memory. Each of the pointer elements addresses objects stored elsewhere, on the heap. The ownership of these objects must be considered when the array is destroyed. If the objects are owned by other components, then it is sufficient to call Close() or Reset() on the array object to clean up the memory associated with it. However, if the objects are owned by the array, they are not destroyed automatically when the array is cleaned up. Thus, as part of cleanup, ResetAndDestroy() must be called to delete the heap object associated with each pointer element in the array.

Exam Essentials

- Demonstrate an understanding of the basics of Symbian OS dynamic arrays (CArrayX and RArray families)
- Understand the different types of Symbian OS dynamic arrays with respect to memory arrangement (flat or segmented), object storage (within array or elsewhere), object length (fixed or variable) and object ownership.
- Recognize the appropriate circumstances for using a segmented-buffer array class rather than a flat array class

8.2 RArray, RPointerArray or CArrayX?

Critical Information

Performance implications of CArrayX dynamic arrays

The original CArrayX classes use the CBufBase base class to access the memory allocated to the array. However, CBufBase works with byte buffers and requires a `TPtr8` object to be constructed for every array access. This results in a performance overhead, even for a simple flat array containing fixed-length elements. Furthermore, for every method which accesses the array there are a minimum of two assertion checks (see Chapter 10) on the incoming parameters, even in release builds.

For example to access a position in a `CArrayFixX` array, `operator[]` calls `CArrayFixBase::At()`, which uses an `__ASSERT_ALWAYS` statement to range-check the index and then calls `CBufFlat::Ptr()`, which also asserts that the position specified lies within the array buffer.

A second issue is that a number of the array-manipulation functions of `CArrayX`, such as `AppendL()`, can leave, for example when there is insufficient memory to resize the array buffer. While it is frequently acceptable to leave under these circumstances, in some cases (such as where the kernel uses the dynamic arrays, or where the array must be called within a function which cannot leave) the leaving functions must be called in a `TRAP` macro to catch any leaves. As described in Chapter 5, the `TRAP` macro has an associated performance overhead.

Prefer the `RArray` and `RPointerArray` classes

The `RArray` and `RPointerArray` classes were added to Symbian OS to provide more efficient, simple flat-memory arrays. When comparing like with like (`RArray` with `CArrayFixFlat`, and `RPointerArray` with `CArrayPtrFlat`), the `RArray` and `RPointerArray` classes have significantly better performance than `CArrayX` classes. Furthermore, they do not need a `TRAP` harness to ensure leave-safe operations when inserting or appending to the array. They can thus be used efficiently both kernel- and user-side.

The `RArray` and `RPointerArray` classes are implemented as R classes, which have a lower overhead than C classes because they do not need the characteristic features of a C class: zero-fill on allocation, a virtual function table pointer and mandatory creation on the heap (for further detail, see Chapter 4).

The searching and ordering functions of the `RArray` classes were also optimized over those of the original classes and were made simpler to use.

The `RArray` or `RPointerArray` classes should be used in preference to the `CArrayFixFlat` and `CArrayPtrFlat` classes whenever the array has the following characteristics:

- The size of an array element is bounded (the current implementation for `RArray` imposes an upper limit of 640 bytes)
- Insertions into the array are relatively infrequent (there is no segmented-memory implementation for `RArray` or `RPointerArray`; both classes use a fixed rather than segmented memory layout).

When a segmented-memory implementation is required, for reasons of efficiency as described in Section 8.1, it may be more appropriate to use the `CArrayX` family of dynamic arrays. For this purpose, `CArrayFixSeg` and `CArrayPtrSeg` are useful alternatives to `RArray` and `RPointerArray`, respectively.

Exceptions

For performance reasons, `RArray` stores objects in the array with word (4-byte) alignment. This means that some member functions do not work when `RArray` is instantiated for classes which are not word-aligned, and an unhandled exception may occur on hardware that enforces strict alignment. The functions affected are:

- The constructor `RArray(TInt, T*, TInt)`
- `Append(const T&)`
- `Insert(const T&, TInt)`
- Operator `[]`, if the returned pointer is used to iterate through the array as for a C array.

Exam Essentials

- Know the reasons for preferring `RArrayX` to `CArrayX`, and the exceptional cases where `CArrayX` classes are a better choice

8.3 Array Granularities

Critical Information

The *capacity* of a dynamic array is the number of elements the array can hold within the space currently allocated to its buffer. When the capacity is filled, the array dynamically resizes itself by reallocating heap memory when the next element is added. The number of additional elements allocated to the buffer is determined by the *granularity*, which is specified at construction time. All dynamic container classes, regardless of whether they have a segmented or flat memory layout, have a granularity for reallocation.

It is important to choose an array granularity consistent with the expected usage pattern of the array. If too small a value is used, an overhead will be incurred for multiple extra allocations when a large number of elements is added to the array. However, if too large a granularity is chosen, the array will waste storage space.

For example, if an array typically holds 8 to 10 objects, then a granularity of 10 would be sensible. A granularity of 100 would be unnecessary. However, if there are usually 11 objects, a granularity of 10 wastes memory for 9 objects unnecessarily. A granularity of 1 would also be foolish, since it would incur multiple reallocations.

Exam Essentials

- Understand the meaning of array granularity and capacity
- Know how to choose the granularity of an array as appropriate to its intended use

8.4 Array Sorting and Searching

Critical Information

CArrayX

For the `CArrayX` classes, an array key can be used to define a property of an array element by which the entire array can be sorted and searched. For example, for an array of task elements, which have an integer priority value and a string name, a key based on the priority may be used to sort the array into priority order. Alternatively, a key based on the name may be used to search for a task of a particular name.

The abstract base class for the array key is `TKey`. The following `TKey`-derived classes implement keys for different types of array:

- `TKeyArrayFix` for arrays of fixed-length elements
- `TKeyArrayVar` for arrays of variable-length elements
- `TKeyArrayPak` for arrays of packed (variable-length) elements.

Accessing an element by key requires the appropriate `TKeyArrayFix`, `TKeyArrayVar` or `TKeyArrayPak` object to be constructed and passed to the `Sort()`, `InsertIsqL()`, `Find()` or `FindIsq()` array-class member function. For example, a search can be made for elements based on the value of a key in one of two ways:

- Sequentially through the array, starting with the first element – performed using the `Find()` member function
- Using a binary-search (binary-chop) technique – performed using the `FindIsq()` member function. This technique assumes that the array elements are sorted in key sequence.

Both functions indicate the success or failure of the search and, if successful, supply the position of the element within the array.

RArray and RPointerArray

RArray classes provide searching and ordering which is more efficient and easier to use than that of their CArrayX counterparts. The objects contained in RArray and RPointerArray may be ordered using a comparator function provided by the element class. That is, the class typically supplies a method which is used to order the objects, and which is passed to the InsertInOrder() or Sort() method by wrapping it in a TLinearOrder<class T> package.

It is also possible to perform lookup operations on the RArray and RPointerArray classes in a similar manner. The RArray classes have several Find() methods, one of which is overloaded to take an object of type TIdentityRelation<class T>. This object packages a function, usually provided by the element class, which determines whether two objects of type T match.

The following code gives an example of a class which uses RArray<class T>. The class, CHerculesTaskManager, is a task manager which stores a set of TTask objects, each of which represents a separate task. The task manager offers an API to allow the tasks to be appended in any order and removed. It also provides functions to list the tasks, either in the order in which they are stored in the array (ListTasksL()) or after first sorting the array into ascending order (ListTasksAscendingL()). The latter function uses the RArray::Sort() method, passing in an object of TLinearOrder<class T>, to sort the contents of the array.

Some of the standard construction code is omitted for clarity in the example. The full code listing can be downloaded from the Symbian Press website for further inspection.

```
const TInt KTaskArrayGranularity = 4;
_LIT8(KTaskEntry, "\n\tTask: ");

enum THerculeanLabours
  {
  ESlayNemeanLion = 1,
  ESlayHydra,
  ... // Other values omitted for clarity
  ECaptureCerberus
  };

enum TTaskManagerPanic
  {
  EInvalidTaskId = 1,
  };

void Panic(TTaskManagerPanic aPanicCode)
  {
  _LIT(KTaskManagerPanic, "TTaskManager");
  User::Panic(KTaskManagerPanic, aPanicCode);
  }
```

```
class TTask
  {
public:
  TTask(THerculeanLabours aLabour);
public:
  static TInt CompareTaskNumbers(const TTask&, const TTask&);
  static TBool MatchTasks(const TTask&, const TTask&);
public:
  TTask(const TTask&);
  TTask& operator=(const TTask&);
public:
  inline const TDesC8& LabourName() const
    {return (iLabourName);};
  inline THerculeanLabours Labour() const
    {return (iLabour);};
private:
  void Initialize();
  TTask(); // Prevent default construction of uninitialized task
private:
  THerculeanLabours iLabour;
  TPtrC8 iLabourName;
  };

class CHerculeanTaskManager : public CBase
  {
public:
  virtual ~CHerculeanTaskManager();
  static CHerculeanTaskManager* NewLC();
public:
  void AppendTaskL(THerculeanLabours aTaskNumber);
  void DeleteTask(THerculeanLabours aTaskNumber);
  void GetTask(THerculeanLabours aTaskNumber, TTask& aTask);
  void ListTasksAscendingL(RBuf8& aTaskList);
  void ListTasksL(RBuf8& aTaskList);
public:
  inline TInt TaskCount() const
    {return (iTaskArray.Count());};
private:
  void SortTasksAscending();
  TInt GetTaskListLength();
private:
  CHerculeanTaskManager();
  void ConstructL();
private:
  RArray<TTask> iTaskArray;
  };

// TTask
TTask::TTask(THerculeanLabours aLabour)
: iLabour(aLabour)
  {// Check that aLabour is a valid value and panic if not
  __ASSERT_ALWAYS(
    ((aLabour>=ESlayNemeanLion)&&(aLabour<=ECaptureCerberus)),
    Panic(EInvalidTaskId));
  Initialize(); // Set the iLabourName
  }
```

```
void TTask::Initialize()
  {
  switch (iLabour)
    {
    case (ESlayNemeanLion):
      {
      _LIT8(KSlayNemeanLion, "SlayNemeanLion");
      iLabourName.Set(KSlayNemeanLion);
      }
    break;
    case (ESlayHydra):
      {
      _LIT8(KSlayHydra, "SlayHydra");
      iLabourName.Set(KSlayHydra);
      }
    break;
    ... // Other values omitted for clarity
    case (ECaptureCerberus):
      {
      _LIT8(KCaptureCerberus, "CaptureCerberus");
      iLabourName.Set(KCaptureCerberus);
      }
    break;
    default:
      ASSERT(EFalse); // Should never get here
    }
  }

// Used by TLinearOrder<class T> in sort operations
// If aTask1.iLabour > aTask2.iLabour return +ve value
// If aTask1.iLabour < aTask2.iLabour return -ve value
// If aTask1.iLabour == aTask2.iLabour return zero
TInt TTask::CompareTaskNumbers(const TTask& aTask1, const TTask& aTask2)
  {
  if (aTask1.iLabour > aTask2.iLabour)
    return (1);
  else if (aTask1.iLabour < aTask2.iLabour)
    return (-1);
  else
    {
    ASSERT(aTask1.iLabour==aTask2.iLabour);
    return (0);
    }
  }

// Used by TIdentityRelation<class T> in search operations
// If tasks match, return ETrue, otherwise EFalse
TBool TTask::MatchTasks(const TTask& aTask1, const TTask& aTask2)
  {
  if (aTask1.iLabour==aTask2.iLabour)
    {
    ASSERT(aTask1.iLabourName.Compare(aTask2.iLabourName)==0);
    return ETrue;
    }

  return (EFalse);
  }
```

```
// CHerculeanTaskManager
CHerculeanTaskManager::CHerculeanTaskManager()
: iTaskArray(KTaskArrayGranularity) // Constructs RArray<TTask>
  {}                                 // with granularity = 4

CHerculeanTaskManager::~CHerculeanTaskManager()
  {// Free memory used by iTaskArray
  iTaskArray.Close();
  }

// Add task to the end of the array
void CHerculeanTaskManager::AppendTaskL(THerculeanLabours aTaskNumber)
  {
  TTask newTask(aTaskNumber);
  User::LeaveIfError(iTaskArray.Append(newTask));
  }

// Deletes all tasks for iLabour==aTaskNumber from iTaskArray
void CHerculeanTaskManager::DeleteTask(THerculeanLabours aTaskNumber)
  {
  TTask tempTask(aTaskNumber);
  TInt foundIndex = iTaskArray.Find(tempTask, TTask::MatchTasks);
  while (foundIndex!=KErrNotFound)
    {
    iTaskArray.Remove(foundIndex);
    foundIndex = iTaskArray.Find(tempTask, TTask::MatchTasks);
    }
  }

// Uses RArray::Find()
void CHerculeanTaskManager::GetTask(THerculeanLabours aTaskNumber,
                                                      TTask& aTask)
  {
  TTask tempTask(aTaskNumber);
  // Creates a TIdentityRelation object implicitly
  // from TTask::MatchTasks
  TInt foundIndex = iTaskArray.Find(tempTask, TTask::MatchTasks);
  aTask = iTaskArray[foundIndex];
  }

// aTaskList is an empty RBuf
// Lists the tasks as they are found in the array
void CHerculeanTaskManager::ListTasksL(RBuf8& aTaskList)
  {
  // Get length of descriptor data required
  TInt listLength = GetTaskListLength();

  aTaskList.CreateL(listLength);

  TInt count = iTaskArray.Count();
  for (TInt index = 0; index<count; index++)
    {
    TTask task = iTaskArray[index];
    aTaskList.Append(KTaskEntry);
    aTaskList.Append(task.LabourName());
    }
  }
```

```
// aTaskList is an empty RBuf
// Sorts the tasks into numerical order starting
// from the lowest value of iLabour
void CHerculeanTaskManager::ListTasksAscendingL(RBuf8& aTaskList)
  {
  SortTasksAscending();
  ListTasksL(aTaskList);
  }

// Uses RArray::Sort()
void CHerculeanTaskManager::SortTasksAscending()
  {
  // Creates a TLinearOrder object implicitly
  // from TTask::CompareTaskNumbers
  iTaskArray.Sort(TTask::CompareTaskNumbers);
  }

// Returns the number of bytes required to list tasks
TInt CHerculeanTaskManager::GetTaskListLength()
  {
  TInt taskEntryLength = KTaskEntry().Length();

  TInt listLength = 0;
  TInt count = iTaskArray.Count();
  for (TInt index = 0; index<count; index++)
    {
    TTask task = iTaskArray[index];
    listLength+=task.LabourName().Length();
    listLength+=taskEntryLength;
    }

  return (listLength);
  }
```

Exam Essentials

- Demonstrate an understanding of how to sort and seek in dynamic arrays
- Recognize that `RArray`, `RPointerArray` and the `CArrayX` family can all be sorted, although the `CArrayX` classes are not as efficient

8.5 TFixedArray

Critical Information

Symbian OS provides a fixed-length array class as an alternative to dynamic arrays. This is useful when the number of elements that will occupy an array is known at compile time.

The `TFixedArray` class wraps the standard fixed-length C++ array (declared using `[]`) and adds range checking to prevent out-of-bounds access. The access is automatically checked in both release and debug builds if the `At()` function is called. Where run-time efficiency is required in production code, operator `[]` can be invoked instead of `At()` so access is bounds-checked in debug builds only.

The checking uses assertion statements (see Chapter 10) and a panic occurs if an attempt is made to use an out-of-range array index. The `TFixedArray` class can be used as a member of a `CBase` class (on the heap) or on the stack, since it is a T class.

Once a `TFixedArray` has been allocated, it cannot be resized dynamically, which is a disadvantage of the class. However, because the allocation has been made, insertion within the bounds of the array is guaranteed to succeed at run-time, which means there is no need to check for out-of-memory errors or leaves on array insertion. Access to the array is fast in release mode.

Besides range-checking, the `TFixedArray` class has some useful additional functions which extend the generic C++ `[]` array. These include:

- `Begin()` and `End()`, for navigating the array
- `Count()`, which returns the number of elements in the array
- `Length()`, which returns the size of an array element in bytes
- `DeleteAll()`, which invokes `delete` on each element of the array
- `Reset()`, which clears the array by filling each element with zeroes.

Besides having to know the size of the array in advance, the main drawbacks to the use of fixed-length arrays are that any additions to an array must occur at the end and that fixed-length arrays do not support ordering and matching.

Exam Essentials

- Recognize that, when a dynamic array is not required, the `TFixedArray` class should be preferred over a C++ array, since it gives the benefit of bounds checking (debug-only or debug and release)

References

[Babin 2005 Chapter 6]
[Stichbury 2004 Chapter 7]

9
Active Objects

Introduction

Active objects are used for event-driven multitasking and are a fundamental part of Symbian OS. This chapter explains why they are so important, and how they are designed for responsive and efficient event handling.

9.1 Event-Driven Multitasking on Symbian OS

Critical Information

Synchronous and asynchronous requests

When program code makes a function call to request a service, the service can be performed either synchronously or asynchronously. When a *synchronous* function is called, it performs a service to completion and returns directly to its caller, usually returning an indication of its success or failure (or leaving, as discussed in Chapter 5). Examples of typical synchronous calls include descriptor-data manipulation methods (see Chapter 7), for example to convert an 8-bit descriptor to 16-bit descriptor, copy descriptor data or convert a descriptor to a number.

An *asynchronous* function submits a request as part of the function call and immediately returns to its caller; the completion of the request occurs some time later. Before the request completes, the caller is free to perform other processing or it may simply wait, which is often referred to as "blocking". Upon completion, the caller receives a signal which indicates the success or failure of the request. This signal is known as an *event*, and the code can be said to be *event-driven*. A timer wait is an example of a typical asynchronous call; another is the `Read()` method on the Symbian OS `RSocket` class (see Chapter 13), which waits to receive data from a remote host.

Symbian OS, like other operating systems, uses event-driven code extensively both at a high level, for example for user interaction, and at a lower, system level, for example for asynchronous communications.

Threads in Symbian OS

In Symbian OS, threads are scheduled *pre-emptively* by the kernel, which runs the highest-priority thread eligible. Each thread may be suspended while waiting for a given event to occur and may resume whenever appropriate. The kernel controls thread scheduling, allowing the threads to share system resources by time-slice division, pre-empting the running of a thread if another, higher-priority thread becomes eligible to run. This constant switching to run the highest-priority ready thread is the basis of pre-emptive multitasking.

A *context switch* occurs when the current thread is suspended (for example, if it becomes blocked, has reached the end of its time-slice, or a higher priority thread becomes ready to run) and another thread is made current by the kernel scheduler. The context switch incurs a run-time overhead in terms of the kernel scheduler and, if the original and replacing threads are executing in different processes, the memory management unit and hardware caches.

Event-driven multitasking

As described above, asynchronously generated events can arise from external sources, such as user input or hardware peripherals that receive incoming data. They can also be generated by software, for example, by timers or completing asynchronous requests. Events are managed by an *event handler*, which, as its name suggests, waits for an event and then handles it.

A high-level example of an event handler is a web-browser application, which waits for user input and responds by submitting requests to receive web pages, which it then displays. The web browser may use a system server, which waits to receive requests from its clients, services them and returns to waiting for another request. The system server submits requests, for example I/O requests, to other servers, which later generate completion events. Each of the software components described is event-driven and needs to be responsive either to user input or to requests from the system (for example, from the communications infrastructure).

In response to an event, the event handler may request another service. This service will later cause another event, or may indicate that the service has completed, which may cause another event in a different part of the system. The operating system must have an efficient event-handling model to handle each event as soon as possible after it occurs and, if more than one event occurs simultaneously, in the most appropriate order. It is particularly important that user-driven events are handled rapidly to give feedback and a good user experience.

Between events, the system should wait in a low-power state. This avoids polling constantly, which can lead to significant power drain and should be avoided on a battery-powered device. Instead, the software should allow the operating system to move to an idle mode while it waits for the next event. On Symbian OS, besides the requirements to be responsive and handle power consumption carefully, it is also important that the memory used by event-handling code is minimized and that processor resources are used efficiently. Active objects achieve these requirements and provide a model for lightweight event-driven multitasking.

Active objects and the active scheduler

Active objects and the active scheduler (collectively known as the "active object framework") are used on Symbian OS to simplify asynchronous programming, making it easy to write code to submit asynchronous requests, manage their completion events and process the results. In general, a Symbian OS application or server will consist of a single main event-handling thread with an associated active scheduler. A number of active objects run in the thread, each with an event-handling function which is called (scheduled) by the active scheduler to handle completed events. Each active object encapsulates a task; it requests an asynchronous service from its service provider and handles the completed event when the active scheduler calls it to do so, communicating with other tasks as necessary.

The active object framework is used to schedule the handling of multiple asynchronous tasks in the same thread. Because the active objects run within the same thread, a switch between them incurs a lower overhead than a thread context switch, This makes it generally the most appropriate choice for event-driven multitasking on Symbian OS. The active objects still run independently of each other, despite existing in the same thread, in much the same way that threads are independent of each other in a process. However, because active objects run in the same thread, memory and objects may be shared more readily.

Thus the active object framework is an example of cooperative (or non-pre-emptive) multitasking: each active object function runs to completion before any other active object in that thread can start to perform an operation. When an active object is handling an event, it cannot be pre-empted by any other running within that thread; however, the thread itself is scheduled pre-emptively, as described above.

Advanced Windows programmers will recognize the pattern of message loop and message dispatch which drives a Win32 application. On Symbian OS, the active scheduler takes the place of the Windows message loop and the event-handling function of an active object acts as the message handler. The event completion processing, performed by the active scheduler, is decoupled from the specifics of event handling, which are performed by individual active objects.

Exceptions

Some events require a response within a guaranteed time, regardless of any other activity in the system. This is called "real-time" event handling. For example, a real-time task may be required to keep the buffer of a sound driver supplied with sound data – a delay in response delays the sound decoding, which results in the sound breaking up. Other typical real-time requirements may be even more strict, say for low-level telephony. The various tasks have different requirements for real-time responses, which can be represented by task priorities. Higher-priority tasks must always be able to pre-empt lower-priority tasks in order to guarantee to meet their real-time requirements. The shorter the response time required, the higher the priority that should be assigned to a task.

However, once an active object is handling an event, it may not be pre-empted by the event handler of another active object within the same thread, which means that they are not suitable for real-time tasks. On Symbian OS, real-time tasks should be implemented using high-priority threads and processes, with the priorities chosen as appropriate for relative real-time requirements.

Exam Essentials

- Demonstrate an understanding of the difference between synchronous and asynchronous requests and be able to differentiate between typical examples of each
- Recognize the typical use of active objects to allow asynchronous tasks to be requested without blocking a thread
- Understand the difference between multitasking using multiple threads and multiple active objects, and why the latter is preferred in Symbian OS code

9.2 Class `CActive`

Critical Information

An active object requests an asynchronous service and handles the resulting completion event some time after the request. It also provides a way to cancel an outstanding request and may provide error handling for exceptional conditions. An active object class must derive directly or indirectly from class `CActive`, defined in `e32base.h`.

`CActive` is an abstract class with two pure virtual functions, `RunL()` and `DoCancel()`, which all inheriting active object classes must define and implement. It also has a `TRequestStatus` member variable which is passed to asynchronous requests to receive the completion result.

Active object class construction

On construction, classes deriving from `CActive` must call the protected constructor of the base class, passing in a parameter to set the priority of the active object. Like threads, all active objects have a priority value to determine how they are scheduled. The reason for this is as follows.

When the asynchronous service associated with the active object completes, it generates an event. The active scheduler detects events, determines which active object is associated with each event, and calls the appropriate active object to handle the event.

While an active object is handling an event, it cannot be pre-empted until the event-handler function has returned back to the active scheduler. It is quite possible that a number of events may complete before control returns to the scheduler. The scheduler must resolve which active object gets to run next; it does this by ordering the active objects using their priority values. If multiple events have occurred before control returns to the scheduler, they are handled sequentially in order of the highest active object priority, rather than in order of completion. Otherwise, an event of low priority that completed just before a more important one would supplant the higher-priority event for an undefined period, depending on how much code it executed to run to completion.

If an active object with a high priority value receives an event while a lower-priority active object is already handling an event, the lower-priority event handler will not be pre-empted. The priority value is only used to determine the order in which event handlers are run.

A set of priority values are defined in the `TPriority` enumeration of class `CActive`. In general, the priority value `CActive::EPriorityStandard` (=0) should be used unless there is good reason to do otherwise.

As an additional part of construction, the active object code should call a static function on the active scheduler, `CActiveScheduler::Add()`. This will add the object to a list, maintained by the active scheduler, of event-handling active objects on that thread. This list is ordered by the active objects' priorities, with the highest-priority objects first.

An active object typically owns a handle to an object to which it issues requests that complete asynchronously, generating an event, such as a timer object of type `RTimer`. This object is generally known as an asynchronous service provider and it may need to be initialized as part of construction. If the initialization can fail, it must be performed as part of the second-phase construction (see Chapter 6).

Submitting an asynchronous service request

An active object class supplies public "request issuer" methods for callers to initiate requests. These will submit requests to the asynchronous service provider associated with the active object, using a well-established pattern, as follows:

1. Check for previous outstanding requests.
   Request methods should check that there is no request already submitted before attempting to submit another. Each active object must only ever have one outstanding request. Depending on the implementation, the code may:

   - Panic if a request has already been issued, if this scenario could only occur because of a programming error
   - Refuse to submit another request, if it is legitimate to attempt to make more than one request
   - Cancel the outstanding request and submit the new one.

2. Issue the request.
   The active object should then issue the request to the service provider, passing in its `iStatus` member variable as the `TRequestStatus&` parameter. The service provider will set this value to `KRequest-Pending` before initiating the asynchronous request.

3. Call `SetActive()` to mark the object as "waiting".
   If the request is submitted successfully, the request method then calls the `SetActive()` method of the `CActive` base class, to indicate to the active scheduler that a request has been submitted and is currently outstanding. This call is not made until after the request has been submitted.

Event handling

Each active object class must implement the pure virtual `RunL()` member method of the `CActive` base class. This is the event handler; when a completion event occurs from the associated asynchronous service provider, the active scheduler selects the active object to handle the event and calls this method.

The `RunL()` function has a slightly misleading name, because the asynchronous function has already run. Perhaps a clearer description would be `HandleEventL()` or `HandleCompletionL()`.

Typical implementations of RunL() determine whether the asynchronous request succeeded by inspecting the completion code, a 32-bit integer value, in the TRequestStatus object (iStatus) of the active object. Depending on the result, RunL() usually either issues another request or notifies other objects in the system of the event's completion; however, the degree of complexity of RunL() code can vary considerably.

Once RunL() is executing, it cannot be pre-empted by other active objects' event handlers. For this reason, the code should complete as quickly as possible so that other events can be handled without delay.

Figure 9.1 illustrates the basic sequence of actions performed when an active object submits a request to an asynchronous service provider that later completes, generating an event which is handled by RunL().

Canceling an outstanding asynchronous request

An active object must be able to cancel any outstanding asynchronous requests it has issued, for example if the application thread in which it is running is about to terminate. The CActive base class implements a Cancel() method, which calls the pure virtual DoCancel() method (which the derived active object class must implement) and waits for the request's early completion. Any implementation of DoCancel() should call the appropriate cancellation method on the asynchronous service provider.

DoCancel() can also include other processing, but should not leave or allocate resources and should not carry out any lengthy operations. It's a good rule to restrict the method to cancellation and any necessary cleanup associated with the request, rather than implementing any sophisticated functionality. This is because a destructor should call Cancel(), as described below, and may already have cleaned up resources that DoCancel() might require.

It isn't necessary to check whether a request is outstanding for the active object before calling Cancel(), because it is safe to do so even if it isn't currently active (that is, awaiting an event).

Error handling

From Symbian OS v6.0 onwards, the CActive class provides a virtual RunError() method which the active scheduler calls if a leave occurs in the RunL() method of the active object. The method takes the leave code as a parameter and returns an error code to indicate whether the leave has been handled. The default implementation does not handle the leave and simply returns the leave code passed to it. If the active object can handle any leaves occurring in RunL() it should do so, by overriding the default implementation of CActive::RunError() to handle the error and return KErrNone to indicate that it has done so.

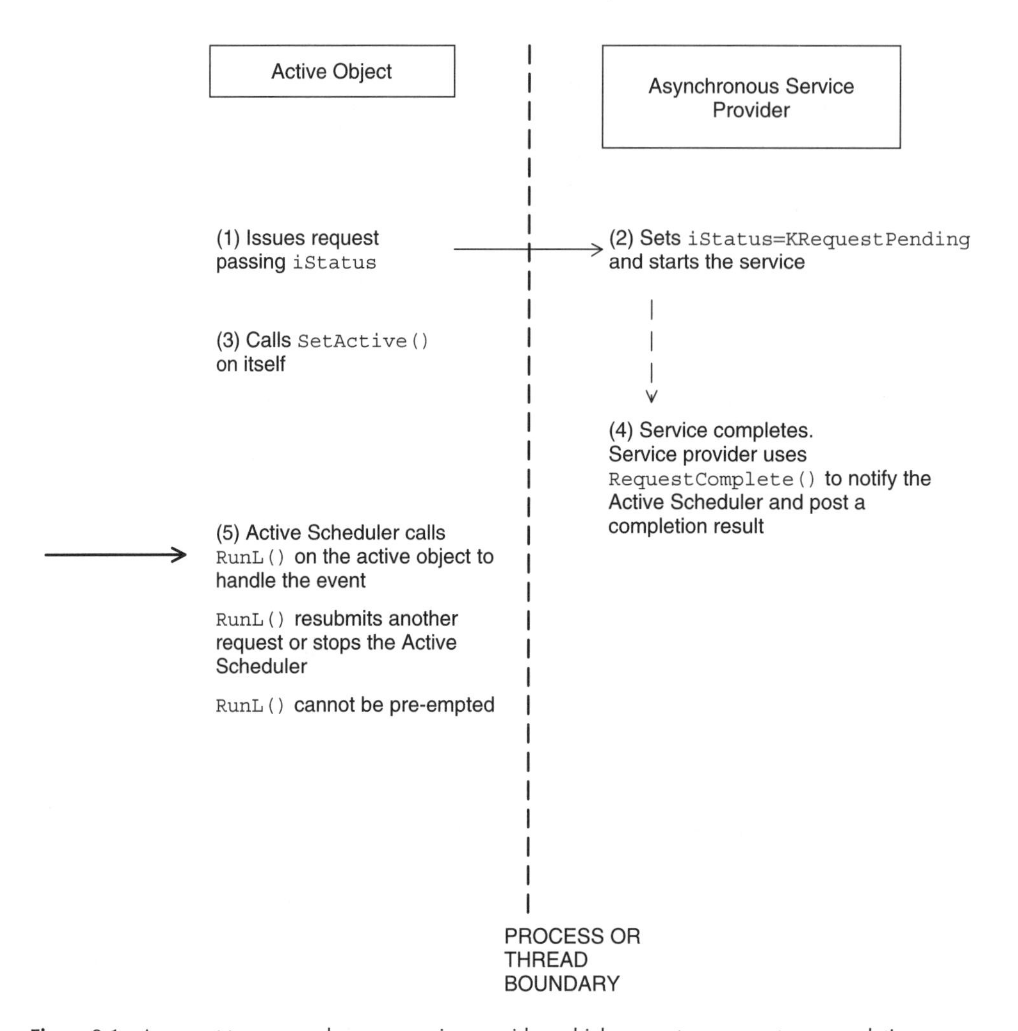

Figure 9.1 A request to an asynchronous service provider, which generates an event on completion

If `RunError()` returns a value other than `KErrNone`, indicating that the leave has yet to be dealt with, the active scheduler calls its own `Error()` function to handle it, as discussed further in Section 9.3. The active scheduler does not have any contextual information about the active object with which to perform error handling, so it is generally preferable to manage error recovery within the `RunError()` method of the associated active object.

Active object class destruction

The destructor of a `CActive`-derived class should always call `Cancel()` to terminate any outstanding requests as part of cleanup code. This should ideally be done before any other resources owned by the active object are destroyed, in case they are used by the service provider or the DoCancel() method. The destructor code should, as usual, free all resources owned by the object, including any handle to the asynchronous service provider.

The CActive base-class destructor is virtual and its implementation checks that the active object is not currently active. It panics if any request is outstanding, that is, if Cancel() has not been called. This catches any programming errors which could lead to the situation where a request completes after the active object to handle it has been destroyed. This would otherwise result in a "stray signal", described in Section 9.6, where the active scheduler cannot locate an active object to handle the event.

Having verified that the active object has no issued requests outstanding, the CActive destructor removes the active object from the active scheduler.

An example of an active object class

The following example illustrates the use of an active object class to wrap an asynchronous service, in this case a timer provided by the RTimer

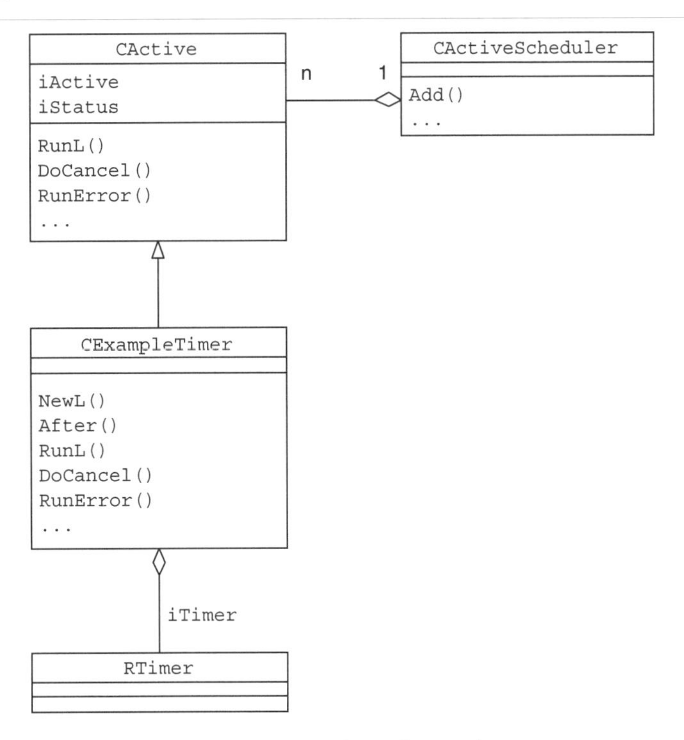

Figure 9.2 CExampleTimer and its relationship with RTimer, CActive and CActiveScheduler

service. In fact, Symbian OS already supplies an abstract active object class, CTimer, which wraps RTimer and can be derived from, so all that needs to be specified is the action to perform when the timer expires. However, the example is used here because it's a straightforward way of showing how to write an active object class. Figure 9.2 shows the classes involved and their relationship with the active scheduler.

When the timer expires, the RunL() event handler checks the active object's iStatus result and leaves if it contains a value other than KErrNone, so that the RunError() method can handle the problem. In this case, the error handling is very simple: the error returned from the request is logged to debug output. This could have been performed in the RunL() method, but has been separated into the RunError() method to demonstrate how to use the active object framework to split error handling from the main logic of the event handler.

If no error occurred, the RunL() event handler logs the timer completion to debug output using RDebug::Print() and resubmits the timer request with the stored interval value. In effect, once the timer request has started, it continues to expire and be resubmitted until it is stopped by a call to the Cancel() method on the active object.

```
class CExampleTimer : public CActive
  {
public:
  ~CExampleTimer();
  static CExampleTimer* NewL();
  void After(TTimeIntervalMicroSeconds32& aInterval);
protected:
  CExampleTimer();
  void ConstructL();
protected:
  virtual void RunL(); // Inherited from CActive
  virtual void DoCancel();
  virtual TInt RunError(TInt aError);
private:
  RTimer iTimer;
  TTimeIntervalMicroSeconds32 iInterval;
  };

CExampleTimer::CExampleTimer() : CActive(EPriorityStandard)
  { CActiveScheduler::Add(this); }

void CExampleTimer::ConstructL()
  {// Create the asynchronous service provider
  User::LeaveIfError(iTimer.CreateLocal());
  }

CExampleTimer* CExampleTimer::NewL() // See Chapter 6
  {...} // Standard 2-phase construction omitted for clarity

CExampleTimer::~CExampleTimer()
  {
```

```
  Cancel();
  iTimer.Close();
  }

void CExampleTimer::After(TTimeIntervalMicroSeconds32& aInterval)
  {// Only allow one timer request to be submitted at a time
   // Caller must call Cancel() before submitting another
  if (IsActive())
    {
    _LIT(KExampleTimerPanic, "CExampleTimer");
    User::Panic(KExampleTimerPanic, KErrInUse));
    }

  iInterval = aInterval;
  iTimer.After(iStatus, aInterval); // Start the RTimer
  SetActive(); // Mark this object active
  }

// Event handler method
void CExampleTimer::RunL()
  {// If an error occurred (admittedly unlikely)
   // deal with the problem in RunError()
  User::LeaveIfError(iStatus.Int());

  // Otherwise, log the timer completion
  _LIT(KTimerExpired, "Timer Expired\n");
  RDebug::Print(KTimerExpired);

  // Resubmit the timer request
  iTimer.After(iStatus, iInterval);
  SetActive();
  }

void CExampleTimer::DoCancel()
  {// Cancel the timer
  iTimer.Cancel();
  }

TInt CExampleTimer::RunError(TInt aError)
  {// Called if RunL() leaves, aError contains the leave code
  _LIT(KErrorLog, "Timer error %d");
  RDebug::Print(KErrorLog, aError); // Logs the error
  return (KErrNone); // Error has been handled
  }
```

Exam Essentials

- Understand the significance of an active object's priority level
- Recognize that the active object event handler method (RunL()) is non-pre-emptive
- Know the inheritance characteristics of active objects, and the functions they are required to implement and override
- Know how to correctly construct, use and destroy an active object

9.3 The Active Scheduler

Critical Information

Creating and installing the active scheduler

Most threads running on Symbian OS have an active scheduler, which is usually created and started implicitly by a framework (for example CONE for the GUI framework). However, server code must create and start an active scheduler explicitly before active objects can be used. Likewise, console-based test code may not use active objects directly itself, but must create an active scheduler in its main thread if it depends on components which use active objects. The code to do this is as follows:

```
CActiveScheduler* scheduler = new(ELeave) CActiveScheduler;
CleanupStack::PushL(scheduler);
CActiveScheduler::Install(scheduler);
```

Starting the active scheduler

Once an active scheduler has been created and installed, its event-processing wait loop is started by a call to the static `CActiveScheduler::Start()` method. The call to `Start()` enters the event-processing loop and does not return until a corresponding call is made to `CActiveScheduler::Stop()`. Thus, before the active scheduler is started, there must be at least one asynchronous request issued via an active object, so that the thread's request semaphore is signaled and the call to `User::WaitForAnyRequest()` completes. If no request is outstanding, the thread simply enters the wait loop and sleeps indefinitely.

The active scheduler wait loop

When an asynchronous request is complete, the asynchronous service provider calls `User::RequestComplete()` if the service provider and requestor are in the same thread, or `RThread::RequestComplete()` if not. It passes to these functions the `TRequestStatus` associated with the request and a completion result, typically one of the standard error codes such as `KErrNone` or `KErrNotFound`, to indicate the success or otherwise of the request. The function called sets the value of `TRequestStatus` to the given error code and generates a completion event in the requesting thread by signaling the thread's request semaphore.

While the request is outstanding, the requesting thread runs in the active scheduler's event-processing loop. When it is not handling other completion events, the active scheduler suspends the thread by calling `User::WaitForAnyRequest()`, which waits for a signal to the thread's request semaphore.

When a signal is received, the active scheduler determines which active object should handle it. It inspects its priority-ordered list of active objects to determine which have outstanding requests (those which after submitting a request called `CActive::SetActive()` to set their `iActive` Boolean to `ETrue`). If an object does indeed have an outstanding request, the active scheduler then inspects its `TRequestStatus` member variable to see if it is set to a value other than `KRequestPending`. If so, this indicates that the active object is associated with a request that has completed and that its event handler code should be called.

Having found a suitable active object, the active scheduler clears the active object's `iActive` Boolean flag and calls its `RunL()` event handler. This method handles the event and may, for example, resubmit a request or generate an event on another object in the system. While this method is running, other events may be generated but `RunL()` is not pre-empted – it runs to completion before the active scheduler resumes control and determines whether any other requests have completed.

Once the `RunL()` call has finished, the active scheduler re-enters the event processing wait loop by issuing another `User::WaitForAnyRequest()` call. This checks the thread's request semaphore and either suspends it (if no other requests have completed in the meantime) or returns immediately (if the semaphore indicates that other events were generated while the previous event handler was running) so that the scheduler can repeat active object lookup and event handling.

Here's some pseudo-code which represents the basic actions of the active scheduler's event-processing loop.

```
EventProcessingLoop()
  {
  // Suspend the thread until an event occurs
  User::WaitForAnyRequest();

  // Thread wakes when the request semaphore is signaled
  // Inspect each active object added to the scheduler,
  // in order of decreasing priority
  // Call the event handler of the first which is active & completed

  FOREVER
    {
    // Get the next active object in the priority queue
    if (activeObject->IsActive())&&
       (activeObject->iStatus!=KRequestPending)
      {// Found an active object ready to handle an event
      // Reset the iActive status to indicate it is not active
      activeObject->iActive = EFalse;

      // Call the active object's event handler in a TRAP
      TRAPD(r, activeObject->RunL());
      if (KErrNone!=r)
        {// event handler left, call RunError() on active object
        r = activeObject->RunError();
```

```
      if (KErrNone!=r) // RunError() didn't handle the error,
        Error(r);      // call CActiveScheduler::Error()
      }
    break; // Event handled, break out of lookup loop and resume
    }
  } // End of FOREVER loop
}
```

The pseudo-code above does not show what happens if the active scheduler inspects all the active objects in its list but cannot find one with an outstanding request and `iStatus` value indicating that it has been completed by the asynchronous service provider. If this occurs, the active scheduler raises a "stray signal" panic (`E32User::CBase 46`). Common reasons for receiving a stray signal are described in Section 9.6.

Stopping the active scheduler

The active scheduler is stopped by a call to `CActiveScheduler::Stop()`. When that enclosing function returns, the outstanding call to `CActiveScheduler::Start()` also returns. Stopping the active scheduler breaks off event handling in the thread, so it should only be called by the main active object controlling the thread.

Extending the active scheduler

`CActiveScheduler` is a concrete class and can be used "as is", but it can also be subclassed. It defines two virtual functions which may be extended: `Error()` and `WaitForAnyRequest()`.

By default, the `WaitForAnyRequest()` function simply calls `User::WaitForAnyRequest()`, but it may be extended, for example to perform some processing before or after the wait. If the function is overridden, it must either call the base-class function or make a call to `User::WaitForAnyRequest()` directly.

As described in Section 9.2, if a leave occurs in a `RunL()` event handler, the active scheduler passes the leave code to the `RunError()` method of the active object. If this method cannot handle the leave, it returns the leave code and the active scheduler passes it to its own `Error()` method. By default, this method just raises a panic (`E32USER-CBASE 47`), but it may be overridden to handle the error, for example by calling an error resolver to obtain the textual description of the error and displaying it to the user or logging it to file.

If active object code is dependent upon particular specializations of the active scheduler, it will not be portable to run in other threads managed by more basic active schedulers. Furthermore, any additional code added to extend the active scheduler should be straightforward and must avoid holding up event handling in the entire thread by performing complex or slow processing.

Exceptions

There are a few threads in Symbian OS which intentionally do not have an active scheduler and thus cannot use active objects or components that use them.

- The Java implementation does not support an active scheduler and native Java methods may not use active objects. It is permissible to make calls in Java code to C++ servers which do use them, since these run in separate threads, each of which has a supporting active scheduler.
- The C Standard Library (`STDLIB`) thread has no active scheduler, thus standard library code cannot use active objects. Functions provided by the Standard Library may however be used in active object code, for example in an initialization or `RunL()` method. The functions should be synchronous and return quickly, as required by all active object implementations.
- OPL does not provide an active scheduler and C++ extensions to OPL (`OPX`s) must not use active objects or any component which uses them. (OPL is an interpreted language generated using an entry-level development tool that enables rapid development of applications.)

Exam Essentials

- Understand the role and characteristics of the active scheduler
- Know that `CActiveScheduler::Start()` should only be called after at least one active object has an outstanding request
- Recognize that a typical reason for a thread to fail to handle events may be that the active scheduler has not been started or has been stopped prematurely
- Understand that `CActiveScheduler` may be sub-classed, and the reasons for creating a derived active scheduler class

9.4 Canceling an Outstanding Request

Critical Information

What happens when `CActive::Cancel()` is called?

When `CActive::Cancel()` is called, it first determines if the active object it has been called on actually has an outstanding request. It does this by checking whether the `iActive` flag is set (by calling

CActive::IsActive()). Section 9.2 describes how this flag is set, by calling CActive::SetActive(); Section 9.3 discusses how it is used by the active scheduler.

If the active object does have an outstanding request, CActive::Cancel() calls DoCancel(), a pure virtual method in CActive, which must be implemented by the derived active object class. When implementing DoCancel(), the code does not need to check if there is an outstanding request, because if there is no outstanding request DoCancel() will not have been called.

The encapsulated asynchronous service provider should provide a method to cancel an outstanding request. DoCancel() calls this method. DoCancel() can include other processing, but it should not carry out any lengthy operations because CActive::Cancel() is a synchronous function which does not return until both DoCancel() has returned and the original asynchronous request has completed. That is, having called DoCancel(), CActive::Cancel() then calls User::WaitForRequest(), passing in a reference to its iStatus member variable. It is blocked until the asynchronous service provider posts a cancellation notification (KErrCancel) into iStatus. Finally, Cancel() resets the iActive member of the active object to reflect that there is no longer an asynchronous request outstanding.

Thus the cancellation event is handled by the Cancel() method of the active object rather than by the active scheduler, and RunL() will not be called.

The Cancel() method of the CActive base class performs all the generic cancellation code. A derived active object class only uses DoCancel() to call the appropriate cancellation function on the asynchronous service provider and to perform any cleanup necessary. DoCancel() should not call User::WaitForRequest(), since this will upset the thread semaphore count.

Calling CActive::Cancel() for active object cleanup

When an active object is about to be destroyed, it must ensure that it is not awaiting completion of a pending request. CActive's destructor removes the active object from the active scheduler, so if any outstanding request it is waiting on completes later, this will generate an event for which there is no associated active object. This results in a stray signal panic (see Section 9.6). To avoid this, the destructor of the CActive base class checks that there is no outstanding request before removing the object from the active scheduler. It will raise an E32USER-CBASE 40 panic if there is, to highlight the problem. For this reason, Cancel() should be called in the destructor of every active object.

CActive::Cancel() invokes the derived class's implementation of DoCancel(), so this method should never contain code which can leave or allocate resources, as it will be called from within the destructor.

Internally, the active object must never call the DoCancel() method directly to cancel a request; it should call CActive::Cancel(), to invoke DoCancel() and handle the resulting cancellation event.

Exam Essentials

- Understand the different paths in code that the active object uses when an asynchronous request completes normally, and as the result of a call to Cancel()

9.5 Background Tasks

Critical Information

Besides encapsulating asynchronous service providers, active objects can also be used to implement long-running tasks which would otherwise need to run in a lower-priority background thread. To be suitable, the task must be divisible into multiple short increments, for example preparing data for printing, performing background recalculations and compacting a database. The increments are performed in the event handler of the active object, which is why they must be short, since RunL() cannot be pre-empted once it is running.

The active object should be assigned a low priority such as CActive::TPriority::EPriorityIdle (=-100), which determines that a task increment only runs when there are no other events to handle, that is, in idle time. If the task consists of a number of different steps, the active object must track the progress as a series of states, implementing it using a state machine.

The active object drives the task by generating its own events to invoke the event handler. That is, instead of calling an asynchronous service provider, it completes itself by calling User::RequestComplete() on its own iStatus object so the active scheduler calls its event handler. In this way it continues to resubmit requests until the entire task is complete.

A typical example is shown in the sample code below; all the relevant methods are shown in the class declarations, but only the implementations relevant to this discussion are given. Error handling is also omitted for clarity. In the example, StartTask(), DoTaskStep() and EndTask() perform small, discrete chunks of the task and can be called directly by the RunL() method of the low-priority active object.

```
class CLongRunningCalculation : public CBase
  {
public:
```

```
  static CLongRunningCalculation* NewL();
  TBool StartTask();  // Initialization before starting the task
  TBool DoTaskStep(); // Performs a short task step
  void EndTask();     // Destroys intermediate data
  ...
  };

TBool CLongRunningCalculation::DoTaskStep()
  {// Do a short task step, returning
  // ETrue if there is more of the task to do
  // EFalse if the task is complete
  ... // Omitted for clarity
  }

class CBackgroundRecalc : public CActive
  {
public:
  ... // NewL(), destructor etc are omitted for clarity
public:
  void PerformRecalculation(TRequestStatus& aStatus);
protected:
  CBackgroundRecalc();
  void ConstructL();
  void Complete();
  virtual void RunL();
  virtual void DoCancel();
private:
  CLongRunningCalculation* iCalc;
  TBool iMoreToDo;
  TRequestStatus* iCallerStatus; // To notify caller on completion
  };

CBackgroundRecalc::CBackgroundRecalc()
: CActive(EPriorityIdle) // Low priority task
  { CActiveScheduler::Add(this); }

// Issues a request to initiate a lengthy task
void CBackgroundRecalc::PerformRecalculation(TRequestStatus& aStatus)
  {
  iCallerStatus = &aStatus;
  *iCallerStatus = KRequestPending;

  _LIT(KExPanic, "CActiveExample");
  __ASSERT_DEBUG(!IsActive(), User::Panic(KExPanic, KErrInUse));

  iMoreToDo = iCalc->StartTask(); // iCalc initializes the task
  Complete(); // Self-completion to generate an event
  }

void CBackgroundRecalc::Complete()
  {// Generates an event on itself by completing on iStatus
  TRequestStatus* status = &iStatus;
  User::RequestComplete(status, KErrNone);
  SetActive();
  }

// Performs the background task in increments
void CBackgroundRecalc::RunL()
```

```
  {// Resubmit request for next increment of the task or stop
  if (!iMoreToDo)
    {// Allow iCalc to cleanup any intermediate data
    iCalc->EndTask();
    // Notify the caller
    User::RequestComplete(iCallerStatus, iStatus.Int());
    }
  else
    {// Submit another request and self-complete to generate event
    iMoreToDo = iCalc->DoTaskStep();
    Complete();
    }
  }

void CBackgroundRecalc::DoCancel()
  {// Give iCalc a chance to perform cleanup
  if (iCalc)
    iCalc->EndTask();

  if (iCallerStatus) // Notify the caller
    User::RequestComplete(iCallerStatus, KErrCancel);
  }
```

Exam Essentials

- Understand how to use an active object to carry out a long-running (or background) task
- Demonstrate an understanding of how self-completion is implemented

9.6 Common Problems

Critical Information

Stray signal panics

The most commonly encountered problem when writing active object code is the infamous "stray signal" panic (`E32USER-CBASE 46`), which occurs when the active scheduler receives a completion event but cannot find an active object to handle it (one which is currently active and has a completed `iStatus` result, indicated by a value other than `KRequestPending`).

Stray signals can arise for the following reasons:

- `CActiveScheduler::Add()` was not called when the active object was constructed
- `SetActive()` was not called following the submission of a request to the asynchronous service provider

- The asynchronous service provider completed the `TRequestStatus` of an active object more than once – either because of a programming error in the asynchronous service provider or because more than one request was submitted simultaneously on the same active object.

Unresponsive event handling

When using active objects for event handling in, for example, a UI thread, event-handler methods must be kept short to keep the UI responsive. No active object should have a monopoly on the active scheduler that prevents other active objects from handling events. Active objects should be "cooperative" and should not:

- Have lengthy `RunL()` or `DoCancel()` methods
- Repeatedly resubmit requests
- Have a higher priority than is necessary.

Blocked thread

A thread can block, and thus prevent an application's UI from remaining responsive, for a variety of reasons including the following:

- A call to `User::After()`, which blocks a thread until the time specified as a parameter has elapsed
- Incorrect use of the active scheduler. Before the active scheduler is started, there must be at least one asynchronous request issued, via an active object, so that the thread's request semaphore is signaled and the call to `User::WaitForAnyRequest()` completes. If no request is outstanding, the thread simply enters the wait loop and sleeps indefinitely (see Section 9.3)
- Use of `User::WaitForRequest()` to wait on an asynchronous request, rather than use of the active object framework.

Exam Essentials

- Know some of the possible causes of stray signal panics, unresponsive event handling and blocked threads

References

[Babin 2005 Chapter 8]
[Harrison 2004 Chapter 1]
[Stichbury 2004 Chapters 8 and 9]

10

System Structure

Introduction

This chapter discusses some of the lower-level features of the Symbian OS platform: DLLs, memory management, threads and processes, interprocess communication (IPC), and panics and assertions.

Before diving into low-level detail, it's worth giving a brief high-level overview. Symbian OS:

- is a multi-tasking operating system, based on open standards, for advanced mobile phones, also known as Smartphones. These phones have a sophisticated graphical user interface (GUI) and a number of built-in applications which use it (for example, messaging and calendar)
- is said to be an ''open'' platform because, in addition to the applications built in by the manufacturer, a user may install others such as games, enterprise applications (for example push e-mail), or various utilities
- is licensed to the world's leading handset manufacturers. At the time of going to press, these are, in alphabetical order: Arima, BenQ, Fujitsu, Lenovo, LG Electronics, Motorola, Mitsubishi, Nokia, Panasonic, Samsung, Sharp and Sony Ericsson
- has a flexible architecture, which allows different user interfaces to run on top of the core operating system. User interfaces designed for Symbian OS include Nokia's S60 and Series 80 platforms, NTT DoCoMo's FOMA user interface and UIQ.

EKA1 and EKA2 refer to different versions of the Symbian OS kernel – the EKA stands for ''EPOC Kernel Architecture'' (Symbian OS was previously known as ''EPOC'', and earlier still, ''EPOC32''). EKA1 is the 32-bit kernel released originally in the Psion Series 5 in 1997. EKA2 was first introduced in Symbian OS version 8.0b, but not shipped in a phone product until version 8.1b, in the Japanese MOAP 2.0 FOMA 902i series phones. EKA2 is the second iteration of Symbian's 32-bit kernel, and is very different internally to EKA1; it offers hard real-time guarantees to kernel and user-mode threads.

10.1 DLLs in Symbian OS

Critical Information

Shared library and polymorphic interface DLLs

Dynamic link libraries, DLLs, are libraries of compiled C++ code that may be loaded into a running process in the context of an existing thread. In Symbian OS there are two main types of DLL: shared library (static-interface) DLLs and polymorphic interface (plug-in) DLLs.

A *shared library* DLL implements library code that may be used by multiple components of any type, that is, other libraries or EXEs. The filename extension of a shared library is `.dll` – examples of this type on Symbian OS are the user library (`EUser.dll`) and the file system library (`EFile.dll`).

A shared library exports API functions according to a module definition (`.def`) file. It may have any number of exported functions, each of which is an entry point into the DLL. A shared library releases a header file (`.h`) for other components to compile against and an import library (`.lib`) to link against, in order to resolve the exported functions.

When executable code that uses the library runs, the Symbian OS loader loads any shared library DLLs that it links to, and any further DLLs that those DLLs require, recursively, until all shared code needed by the executable is loaded.

The second type of DLL, a *polymorphic interface* DLL, implements an abstract interface which is usually defined separately, for example by a framework. It may have a `.dll` filename extension, but it often uses a different extension to identify the nature of the DLL further, for example `.fsy` for a file system plug-in (see Chapter 12), or `.prt` for a protocol module plug-in (see Chapter 13).

Polymorphic DLLs have a single entry-point "gate" or "factory" function, which instantiates the concrete class that implements the interface. Polymorphic interface DLLs are often used to provide a range of different implementations (plug-ins) of a single consistent interface. They are loaded dynamically, typically by a framework.

From Symbian OS v7.0 onward, the most common type of plug-ins are ECOM plug-ins. ECOM is a generic framework for specifying interfaces, and for finding and loading those plug-ins which implement them. Many Symbian OS frameworks, such as the recognizer framework, discussed in

Section 10.6 now require their plug-ins to be written as ECOM plug-ins, rather than as a "proprietary" type of polymorphic interface DLL. The use of ECOM allows each framework to delegate the finding and loading of suitable plug-ins to ECOM, rather than performing that task itself.

UIDs used by DLLs

A Symbian OS UID is a 32-bit, globally *unique identifier* value used to identify a file type, both for running executable code and for associating data files with the appropriate application. Symbian OS uses a combination of up to three UIDs to uniquely identify a binary executable. For DLLs, the three UID values are used as follows:

- UID1 is a system-level identifier which distinguishes between EXEs and DLLs. This value is never stated explicitly but is determined by the Symbian build tools from the `targettype` specified in the `MMP` file. For shared libraries, the `targettype` specified should be `DLL` (UID1 = `KDynamicLibraryUid = 0x10000079`), while for polymorphic ECOM plug-in DLLs, the `targettype` is `PLUGIN` (or `ECOMIIC` for versions of Symbian OS earlier than v9.0). Other polymorphic plug-in DLL target types, which are not ECOM plug-ins, include `FSY` (file system plug-in) and `PRT` (protocol module plug-ins). The `targettype` keyword and the build tools are discussed in more detail in Chapter 14
- UID2 distinguishes between shared library DLLs (`KSharedLibraryUid = 0x1000008d`) and polymorphic interface DLLs, which vary because they are assigned a UID2 value specific to their type (for example, the socket server protocol module UID2 value is `0x1000004A`)
- UID3 identifies a component uniquely. In order to ensure that each binary that needs a distinguishing UID is assigned a genuinely unique value, Symbian manages UID allocation through a central database and developers must be registered with Symbian Signed to request UIDs (see Chapter 15).

For EXEs, the UID1 value is set by the choice of `targettype EXE` to be `KExecutableImageUid` (`0x1000007a`). UID2 is not relevant for an EXE and can be left unspecified or set explicitly to `KNullUid` (= 0). UID3 can be left unspecified too but, on Symbian OS v9 and beyond, it should usually be set to a unique value to act as the secure identifier for the binary (see Chapter 15).

Exporting functions from a DLL

A shared library DLL provides access to its APIs by exporting its functions so that separate executables (that is, DLL or EXE code compiled into a separate binary component) can call them. This makes the functions "public" to other modules by creating a .lib file, which contains the export table to be linked against by the calling code.

On Symbian OS, every function to be exported should be marked in the class definition in the header file with the macro IMPORT_C. The client code will include the header file, so they are effectively "importing" each function into their code module when they call it. The corresponding function should be prefixed with the EXPORT_C macro in the .cpp file which implements it. For example:

```
class CMyExample : public CSomeBase
  {
public:
  IMPORT_C static CMyExample* NewL();
public:
  IMPORT_C void Foo();
  ...
  };

EXPORT_C CMyExample* CMyExample::NewL()
  {...}

EXPORT_C void CMyExample::Foo()
  {...}
```

The rules as to which functions should be exported are as follows:

- Inline functions must never be exported, because there's no need to do so. The IMPORT_C and EXPORT_C macros add functions to the export table to make them accessible to components linking against the library. However, the code of an inline function is, by definition, already accessible to callers, since it is declared within the header file (and the compiler interprets the inline directive by adding the code directly into the client code wherever it calls it)
- Only functions which need to be used outside a DLL should be exported. The use of IMPORT_C and EXPORT_C adds an entry to the export table in the module definition (.def) file. If the function is private to the class and can never be accessed by client code, exporting it merely adds it to the export table unnecessarily
- All virtual functions, whether public, protected or private, should be exported, since they may be re-implemented by a derived class in another code module. Any class which has virtual functions must also export a constructor, even if it is empty, so that the virtual function table can be correctly generated by access to the base-class constructor.

Lookup by ordinal and by name

On Symbian OS, the size of DLL program code is optimized to save ROM and RAM space. In most operating systems, to load a dynamic library, the entry points of a DLL can either be identified by string-matching their name (lookup by name) or by the order in which they are exported in the module definition file (lookup by ordinal).

Symbian OS does not offer lookup by name, because this adds an overhead to the size of the DLL (storing the names of all the functions exported from the library is wasteful of limited ROM and RAM space). Instead, Symbian OS only uses link by ordinal. This has significant implications for binary compatibility; ordinals must not be changed between one release of a DLL and another. For example, code which links against a library and uses an exported function with a specific ordinal number in an early version of the library will not be able to call that function in a newer version of the library if the ordinal number is changed.

Binary compatibility is discussed further in Chapter 16.

Exceptions

The one type of virtual function which should not be exported from a DLL is a pure virtual function, because there is generally no implementation code for a pure virtual function, so there is no code to export.

Exam Essentials

- Know and understand the characteristics of polymorphic interface and shared library (static) DLLs
- Know that UID2 values are used to distinguish between static and polymorphic DLLs, and between plug-in types
- For a shared library, understand which functions must be exported if other binary components are to be able to access them
- Know that Symbian OS does not allow library lookup by name but only by ordinal

10.2 Writable Static Data

Critical Information

Versions of Symbian OS which support writable static data

Symbian OS supports global writable static data in EXEs on all versions and handsets.

However, writable static data cannot be used in DLLs built for target hardware for Symbian OS versions 8.1a, 8.0a or earlier, that is, those versions of Symbian OS which contain EKA1. This is because, on that version of the platform, DLLs have separate areas for program code and read-only data but do not have an area for writable data.

Symbian OS versions 8.0b, 8.1b, 9.0 and beyond do now support the use of writable static data in DLLs, but it is still not recommended, because it is expensive in terms of memory usage and has limited support in the Symbian OS Emulator. Even on Symbian OS versions where it is supported, Symbian recommends that it only be used as a last resort, for example when porting code written for other platforms which uses writable static data heavily.

On EKA1, which did not support writable static data in DLLs, all GUI applications were built as such, which meant that no application code could use writable static or global data. On EKA2, applications are now built as EXEs, so this is no longer an issue – modifiable global or static data has always been allowed in EXEs.

Symbian OS platform version	Writable static data in DLLs built for hardware	Application binary type
v6.1 – v8.0a (inclusive), v8.1a	Not supported on hardware builds (compilation will fail)	DLL – no writable static data allowed
v8.0b, v8.1b, v9.0 and beyond	Supported but not recommended – limited emulator support and inefficient in terms of memory usage	EXE – writable static data can be used

In order to enable global writable static data on EKA2, the `EPOCALLOWDLLDATA` keyword must be added to the `MMP` file of a DLL (see Chapter 14). Where this is not used, and on EKA1 versions of the Symbian OS, the PETRAN build tool will return an error when the DLL code is built for the phone hardware.

Workarounds to avoid writable static data

1. Thread-local storage.
   One workaround used to replace writable static data is called *thread-local storage* (TLS). This can be accessed through class `Dll` on pre-8.1b versions of Symbian OS, and through class `UserSvr` for version 8.1b and version 9.0.

Thread-local storage is simply a 32-bit pointer, specific to each thread, that can be used to refer to an object which simulates global writable static data. All the global data must be grouped within this single object, which is allocated on the heap on creation of the thread. The pointer to the object is saved to the thread-local storage pointer, using `Dll::SetTls()` or `UserSvr::DllSetTls()`. To access the global data, the code calls `Dll::Tls()` or `UserSvr::DllTls()`. On destruction of the thread, the data is destroyed too.

2. Client–server framework.
   Symbian OS supports writable global static data in EXEs. A common porting strategy is to wrap the code in a Symbian server (which is an EXE), and expose its API as a client interface.

3. Embed global variables into classes.
   With relatively small amounts of code, it may be possible to move most global data inside classes. The data can then be passed as function parameters between objects and functions, where necessary.

Writable static data defined

Global writable static data is any per-process modifiable variable which exists for the lifetime of the process. In practice, this means any globally scoped data declared outside of a function, struct or class, as well as function-scoped static variables.

The only global data that can be used within DLLs is constant global data of the built-in types, or of a class with no constructor. These definitions are acceptable:

```
static const TUid KUidFooDll = { 0xF000C001 };
static const TInt KMinimumPasswordLength = 6;
```

The following definitions cannot be used because they have non-trivial class constructors, which require the objects to be constructed at run-time.

```
static const TPoint KGlobalStartingPoint(50, 50);
static const TChar KExclamation('!');

// The following literal type is deprecated (see Chapter 7)
static const TPtrC KDefaultInput =_L("");
```

This means that, although the memory for the object is pre-allocated in code, it doesn't actually become initialized and constant until after the constructor has run. Thus, at build time, each constitutes a non-constant global object and causes the build to fail for phone hardware.

The following object is also non-constant because, although the data pointed to by `ptr` is constant, the pointer itself is not constant:

```
// Writable static data!
static const TText* ptr = (const TText*)"data";
```

This can be corrected by making the pointer constant (see Chapter 1):

```
static const TText* const ptr = (const TText*)"data";
```

Exceptions

On EKA1, the emulator can use the underlying Windows DLL mechanism to provide per-process DLL data. If non-constant global data is used inadvertently, it will go undetected in emulator builds and will only fail when the PETRAN tool encounters it in the hardware platform build.

Exam Essentials

- Recognize that writable static data is not allowed in DLLs on EKA1 and discouraged on EKA2
- Know the basic porting strategies for removing writable static data from DLLs

10.3 Executables in ROM and RAM

Critical Information

EXEs in ROM and RAM

On target hardware, executable code can either be built onto the phone in read-only memory (ROM) when the phone is in the factory, or can be later installed on the phone, either into the phone's internal memory or onto removable storage media such as a memory stick or MMC.

As a generalization, ROM-based EXEs can be thought of as executing directly in place from the ROM. This means that program code and read-only data (such as literal descriptors) are read directly from the ROM, and the component is only allocated a separate data area in RAM for its read/write data.

If an EXE is installed, rather than built into the ROM, it executes entirely from RAM and has an area allocated for program code and read-only static data, and a separate area for read/write static data. If a second copy of the EXE is launched, the read-only area is shared, and only a new area of read/write data is allocated.

DLLs in ROM and RAM

DLLs in ROM are not actually loaded into memory either, but execute in place in ROM at their fixed address. DLLs running from RAM are loaded at a particular address and reference counted so they are unloaded only when no longer being used by any component. When a DLL runs from RAM, the address at which the executable code is located is determined only at load time. Loading a DLL from RAM is different from simply storing it on the internal (RAM) drive, because Symbian OS copies it into the area of RAM reserved for program code and prepares it for execution by fixing up the relocation information.

The relocation information to navigate the code of the DLL must be retained for use in RAM. However, DLLs that execute from ROM are already fixed at an address and do not need to be relocated. Thus, to compact the DLL in order to occupy less ROM space, Symbian OS tools strip the relocation information out when a ROM is built. This does mean, however, that a DLL cannot be copied from the ROM, stored in RAM and run from there.

For both types of DLL, static and polymorphic, the code section is shared. This means that, if multiple threads or processes use a DLL simultaneously, the same copy of program code is accessed at the same location in memory. Subsequently loaded processes or libraries that wish to use it are "fixed up" by the DLL loader to use that copy.

Exam Essentials

- Recognize the correctness of basic statements about Symbian OS execution of DLLs and EXEs in ROM and RAM

10.4 Threads and Processes

Critical Information

Threads

Threads form the basis of multitasking and allow multiple sequences of code to execute simultaneously. It is possible to create multiple threads in a Symbian OS application for parallel execution, but in many cases it is more appropriate to use active objects, which are optimized for event-driven multi-tasking on Symbian OS (see Chapter 9).

On Symbian OS, the class used to manipulate threads is `RThread`, an R class (see Chapter 4). An object of type `RThread` represents a handle to a thread, because the thread itself is a kernel object. The base class of `RThread` is `RHandleBase`, which encapsulates the behavior of a generic handle and is used as a base class throughout Symbian OS to identify a handle to another object, often a kernel object.

Class `RThread` defines several functions for thread creation. Threads are not contained in separate executable files but execute within a parent process executable, although each thread has an independent execution stream.

Each thread-creation function takes a descriptor representing a unique name for the new thread, a pointer to a function in which thread execution starts, a pointer to data to be passed to that function, and a value for the stack size of the thread, which defaults to 8 KB. The `Create()` function is overloaded to offer various options associated with the thread heap, such as its maximum and minimum size and whether it shares the creating thread's heap or uses a specific heap within the process in which it runs.

By default, each Symbian OS thread has its own independent heap as well as its own stack. The size of the stack is limited to the size set in `RThread::Create()`, but the heap can grow from its minimum size up to a maximum size. Where the thread has its own heap, the stack and the heap are located in the same chunk of memory.

When the thread is created, it is assigned a unique thread identity, which is returned by the `Id()` function of `RThread` as a `TThreadId` object. If the identity of an existing thread is known, it can be passed to `RThread::Open()` to open a handle to that thread. Alternatively, the unique name of a thread can be passed to open a handle to it.

A thread is created in the suspended state and its execution started by a call to `RThread::Resume()`. On Symbian OS, threads are preemptively scheduled and the currently running thread is the highest-priority thread ready to run. If there are two or more threads with equal priority, they are time-sliced on a round-robin basis. The priority of a thread is a number: the higher the value, the higher the priority.

A running thread can be removed from the scheduler's ready-to-run queue by a call to `Suspend()` on its thread handle. It still exists, however, and can be scheduled to run again by another call to `Resume()`. A thread can be ended permanently by a call to `Kill()` or `Terminate()`, both of which take an integer parameter representing the exit reason. These methods should be used to stop a thread normally, while `Panic()` is used for stopping the thread to highlight a programming error (see Section 10.7).

On EKA1, a thread must call `SetProtected()` to prevent other threads from acquiring a handle to it and stopping it by a call to `Suspend()`, `Panic()`, `Kill()` or `Terminate()`. On EKA2, the security model ensures that a thread is always protected from threads running in other processes, and the redundant `SetProtected()` method has been removed. That is, the default protection for EKA2 ensures that it is no longer possible for a thread to stop another thread in a different process by calling `Suspend()`, `Terminate()`, `Kill()` or `Panic()` on it. The functions are retained in EKA2 because a thread can still call the various termination functions on itself or other threads in the same process. It is also still possible for a server to panic a misbehaving client thread, if necessary, by calling `RMessagePtr2::Panic()`.

If the main thread in a process is ended by any of these methods, the process also terminates. However, if a secondary thread (that is, one created from within the running process by a call to `RThread::Create()`) terminates, the process itself does not stop running.

It is also possible to receive notification when a thread dies. A call to `RThread::Logon()` on a valid thread handle, passing in a `TRequestStatus` reference, submits a request for notification when that thread terminates. The request completes when the thread terminates, and receives the value with which the thread ended, or `KErrCancel` if the notification request was cancelled by a call to `RThread::LogonCancel()`. The thread handle class also provides functions `RThread::ExitType()`, `RThread::ExitReason()` and `RThread::ExitCategory()` to give full details of the associated thread's end state.

A rendezvous request can also be created with another thread by calling the asynchronous `RThread::Rendezvous()` function on the thread handle. The asynchronous request completes in any of the following ways:

- when the thread in question next calls `RThread::Rendezvous(TInt aReason)`, which completes the asynchronous request with `aReason`
- if the outstanding request is cancelled by a call to `RThread::RendezvousCancel()`, whereupon the asynchronous request completes with `KErrCancel`
- if the thread exits or panics, which completes the asynchronous request with the thread exit reason value (that is, the value passed to the `Kill()`, `Exit()` or `Panic()` function which caused the termination of the thread).

Besides the use of RThread::Rendezvous(), Symbian OS provides several classes representing kernel objects for thread synchronization:

- a *semaphore* can be used either for sending a signal from one thread to another, or for protecting a shared resource from being accessed by multiple threads at the same time. On Symbian OS, a semaphore is created and accessed with a handle class called `RSemaphore`. A global semaphore can be created, opened and used by any process in the system, while a local semaphore can be restricted to all threads within a single process. Semaphores can be used to limit concurrent access to a shared resource, either to a single thread at a time, or allowing multiple accesses up to a specified limit

- a *mutex* is used to protect a shared resource so that it can only be accessed by one thread at a time. On Symbian OS, the `RMutex` class is used to create and access global and local mutexes
- a *critical section* is a region of code that should not be entered simultaneously by multiple threads. An example is code that manipulates global static data, since it could cause problems if multiple threads change the data simultaneously. Symbian OS provides the `RCriticalSection` class that allows only one thread within the process into the controlled section, forcing other threads attempting to gain access to that critical section to wait until the first thread has exited from the critical section. `RCriticalSection` objects are always local to a process, and a critical section cannot be used to control access to a resource shared by threads across different processes – a mutex or semaphore should be used instead.

Processes

A Symbian OS process is an executable that has its own data area, stack and heap; by default a process is given 8 KB of stack and 1 MB of heap. Many processes can be active on Symbian OS at once, including multiple instances of the same process. Processes have private address spaces and a user-side process cannot directly access memory belonging to another user-side process.

By default, a process contains a single execution thread, the main thread, but additional threads can be created as described above. A context switch to a process occurs whenever one of the threads in that process is scheduled to run and becomes active. Switching between threads in different processes is more "expensive" than switching between threads within the same process; this is because a process switch requires that the data areas of the two processes be remapped by the memory management unit (MMU).

On Symbian OS, the class used to manipulate processes is `RProcess`. In much the same way as described above for `RThread`, the `RProcess::Create()` function can be used to start a new, named process, and the `RProcess::Open()` function can be used to open a handle to a process identified by name or process identity (`TProcessId`). Assorted functions to stop the process are also similar. A `Resume()` function is also provided by `RProcess`, which marks the first thread in the process as eligible for execution. However, there is no `RProcess::Suspend()`

function because processes are not scheduled; threads form the basic unit of execution and run inside the protected address space of a process.

On Windows, the emulator runs within a single Win32 process, `EPOC.exe`, and each Symbian OS process runs as a separate thread inside it. On EKA1, the emulation of processes on Windows is incomplete and `RProcess::Create()` returns `KErrNotFound`. The EKA2 release has removed this inconvenience: while Symbian OS still runs in a single process on Windows, the emulation is enhanced and `RProcess::Create()` translates to creation of a new Win32 thread within `EPOC.exe`.

Exam Essentials

- Recognize the correctness of basic statements about threads and processes on Symbian OS
- Recognize the role and the characteristics of the synchronization primitives `RMutex`, `RCriticalSection` and `RSemaphore`

10.5 Inter-Process Communication (IPC)

Critical Information

Client–server framework

Chapter 11 describes a common form of inter-process communication (IPC) on Symbian OS: the client–server framework.

Clients connect to servers and establish a session for all further communication, which consists of client requests and server responses, mediated by the kernel. Session-based communication ensures that all clients will be notified in the case of an error or shutdown of a server, and all server resources will be cleaned up if an error occurs, or when a client disconnects or dies.

This type of communication paradigm is ideal when many clients need reliable concurrent access to a service or shared resource. In that case the server serializes and mediates access to the service accordingly. However, there are some limitations:

- clients must know which server provides the service they need
- a permanent session must be maintained between client and server
- it is not really suitable for event multicasting (server-initiated "broadcast" to multiple clients).

In order to overcome such limitations and enrich the IPC mechanisms, Symbian OS version 8.0 was extended to offer additional IPC mechanisms:

publish and subscribe, message queues and shared buffer I/O. Publish and subscribe and message queues are described in more detail below. Shared buffer I/O is not discussed because it is intended primarily for device driver developers (it is used to allow a device driver and its clients to access the same memory area without copying, even during interrupt handling).

Publish and subscribe

The publish and subscribe mechanism was created to provide asynchronous multicast event notification, and to allow for connectionless communication between threads.

Publish and subscribe provides a means to define and publish changes to system-wide global variables known as "properties". Changes to the properties can be communicated ("published") to more than one interested ("subscribed") peer asynchronously. Publishers and subscribers can dynamically join and leave without any connection set-up or teardown.

Properties are data values, uniquely identified by a 64-bit integer, which is the only information that must be shared between a publisher and a subscriber (typically through a common header file). There is no need to provide interface classes or functions for a property. Subscribers do not need to know which component is publishing to a property, they only need to know about the publish and subscribe API, and the identity of the property of interest to them.

The Symbian OS publish and subscribe API is supplied by the `RProperty` class. The identity of a property is composed of two parts:

- a category (defined by a standard UID) which specifies the category to which the property belongs
- a key, which uniquely identifies a property within a particular category. Its value depends on how keys within the category are enumerated.

Once identified, a property holds a single data variable which may be either a 32-bit integer, a byte array (a descriptor) of up to 512 bytes in length, Unicode text (also up to 512 bytes in size), or even large byte arrays of up to 65 536 bytes.

A thread may take the role of either the publisher or the subscriber, and any party interested in a property can be the one to define it, by calling `RProperty::Define()` to create the variable and specify its type and access controls. Once a property has been defined, it will persist in the kernel until it is deleted explicitly or the system reboots. The property's lifetime is not linked to that of the defining thread or process.

Properties can be published or retrieved either using a previously attached handle or by specifying the property's identity for each call.

On EKA2, the benefit of the former method is that it has a deterministic bounded execution time, making it suitable for high-priority, real-time tasks.

A property is published by calling `RProperty::Set()`. This writes a new value atomically to the property, thus ensuring that access by multiple threads is handled correctly. When a property is published, all outstanding subscriptions are completed, even if the value is actually unchanged. This allows the property to be used as a simple broadcast notification.

To subscribe to a property, a client must register interest by attaching to it and calling the asynchronous `RProperty::Subscribe()` method. Notification happens in the following stages:

1. A client registers its interest in the property by attaching to it (`RProperty::Attach()`) and calling `Subscribe()` on the resulting handle, passing in a `TRequestStatus` reference
2. Upon publication of a new value, the client gets notified, via a signal to the `TRequestStatus` object to complete the `Subscribe()` request
3. The client retrieves the value of the updated property by calling `RProperty::Get()`
4. The client can re-submit a request for notification of changes to the property by calling `Subscribe()` again.

It is not necessary for a property to be defined before it is accessed (lazy definition) so it is not a programming error for a property to be published before it has been defined. This is known as speculative publishing. Attaching to an undefined property is not necessarily an error. Likewise, a `Subscribe()` request on an undefined property will not complete until either the property is defined and published, or the subscriber unsubscribes by canceling the request, using `RProperty::Cancel()`.

Publish and subscribe should be used when a component needs to supply or consume timely and transient information to or from an unknown number and type of interested parties, while remaining decoupled from them. A typical example is the notification of a change to the device's radio states; for example flight-mode, Bluetooth radio on/off, WiFi on/off, etc.

Publish and subscribe and platform security

On the secure platform of Symbian OS v9, to ensure that processes are partitioned so that one process cannot interfere with the property of another process, the category UID of the property should match the secure identifier of the defining process. Alternatively, the process calling

`RProperty::Define()` must have `WriteDeviceData` capability (see Chapter 15). Properties must also be defined with security policies (using `TSecurityPolicy` objects):

- to specify the capabilities and/or vendor identifier and/or secure identifier required for processes to publish the property value
- to specify the capabilities and/or vendor identifier and/or secure identifier required for processes to subscribe to the property.

For example, before accepting a subscription to a property, the security policy defined when the property was created is checked, and the subscription request completes with `KErrPermissionDenied` if the check fails. Platform Security is described in more detail in Chapter 15.

Message queues

In contrast to the connection-oriented nature of client–server IPC, message queues (`RMsgQueue`) offer a peer-to-peer, many-to-many communication mechanism. Message queues provide a way to send data (messages) to interested parties without needing to know whether any thread is listening or the identity of a recipient; in effect, messages are sent to the queue rather than to any specific recipient. A single queue can be shared by many readers and writers.

A message is an object that is placed into a queue for delivery to recipients. A queue is normally created for messages of a given type. This means that a queue is created to deal with messages of a defined (fixed) length, which must be a multiple of four bytes. The size of a queue (the maximum number of messages, or slots, it can contain) is also fixed when the queue is created. The maximum size of message for which a queue is created, and the maximum size of the queue are limited only by system resources.

A message queue allows two or more threads to communicate without setting up a connection to each other. It is a mechanism for passing data:

- between threads that run in separate processes (using a global queue which is named and visible to other processes)
- between threads within a process (using a local queue which is not visible to other processes). The messages can point to memory mapped to that process and can be used for passing descriptors and pointers between threads.

Message queues allow for "fire-and-forget" IPC from senders to recipients and lend themselves well to event notification.

While publish and subscribe is good for notification of state changes which are inherently transient, message queues are useful for allowing information to be communicated beyond the lifetime of the sender. For example, a central logging subsystem can use a message queue to receive messages from numerous threads that may or may not still be running at the point the messages are read and processed.

However, neither messages nor queues are persistent; they are cleaned up when the last handle to the queue is closed.

Exam Essentials

- Recognize the preferred mechanisms for IPC on Symbian OS (client–server, publish and subscribe and message queues), and demonstrate awareness of which mechanism is most appropriate for given scenarios
- Understand the use of publish and subscribe to retrieve and subscribe to changes in system-wide properties, including the role of platform security in protecting properties against malicious manipulation

10.6 Recognizers

Critical Information

Recognizers are a good example of the use of framework plug-in DLLs. The framework which loads the recognizers is provided by the application architecture server (Apparc). Apparc implemented its own custom loading of recognizer plug-ins up to v9.1; in later releases it has been modified to use ECOM. Apparc can load any number of recognizer plug-in DLLs.

When a file in the file system needs to be associated with an application, Apparc opens the file and reads some data from the start of it into a buffer. It then calls `DoRecognizeL()` on each recognizer in the system in turn, passing in the data it read into the buffer. If a plug-in "recognizes" it, it returns its data type (MIME type). Recognizers do not handle the data; they just try to identify its type so that the data can be passed to the application that can best use it.

The plug-in recognizer architecture allows developers to create additional data recognizers and add them to the system by installing them. All data recognizers must implement the polymorphic interface defined by `CApaDataRecognizerType`, which has three virtual functions:

- `DoRecognizeL()` – performs data recognition. All recognizers must implement this method although it is not pure virtual. Each implementation should set a value to indicate the MIME type it considers the data to belong to, and another value to indicate a level of confidence, ranging from `ECertain` (the data is definitely of a specific data type) to `ENotRecognized` (the data is not recognized).

- `SupportedDataTypeL()` – returns the MIME types that the recognizer is capable of recognizing. This pure virtual function must be implemented by all recognizer plug-ins. Each recognizer's implementation of `SupportedDataTypeL()` is called by the recognizer framework after all the recognizers in the system have been loaded, to build up a list of all the types the system can recognize.
- `PreferredBufSize()` – specifies the size in bytes of the buffer passed to `DoRecognizeL()` that the recognizer needs to work with. This function is not pure virtual, but must be implemented.

Exam Essentials

- Recognize correct statements about the role of recognizers in Symbian OS

10.7 Panics and Assertions

Critical Information

Panics

On Symbian OS, when a thread is panicked, it stops running. Panics are used to highlight a programming error in the most noticeable way, stopping the thread to ensure that the code is fixed, rather than potentially causing serious problems by continuing to run. There is no recovery from a panic. Unlike a leave, a panic can't be trapped; a panic is terminal.

If a panic occurs in the main thread of a process, the entire process in which the thread runs will terminate. If a panic occurs in a secondary thread, it is only that thread which closes. However, if a thread is deemed to be a system thread, that is essential for the system to run, a panic in that thread will reboot the phone. This is very rare since the code running in system threads on Symbian OS is mature and well-tested.

On phone hardware and in release builds on the Windows emulator, the end result of a panic is either a reboot or an "Application closed" message box. In debug emulator builds, a panic can be set to break into the debugger – this is known as "just-in-time" debugging. The debugger can then be used to look through the call stack to see where the panic arose and examine the state of appropriate objects and variables.

A call to the static function `User::Panic()` panics the currently running thread. On EKA2, as described in Section 10.4, a thread may panic any other thread in the same process by acquiring an `RThread` handle and using it to call `RThread::Panic()`. On EKA1, this function could be used to panic any unprotected thread in the system, but this was deemed insecure for EKA2. The only occasion where a thread running inside a user process can panic another thread in a different process is for a server thread to panic a badly-behaved client by using the `RMessagePtr2::Panic()` method.

Both `User::Panic()` and `RThread::Panic()` take two parameters: a panic category string and an integer error code, which can be any value, positive, zero or negative. Even without breaking into the debugger, these values should still be sufficient for a developer to determine the cause of a panic. The panic string should be short and descriptive for a programmer rather than for a user, since the user should **never** see them. This is because panics should only be used as a means to eliminate programming errors during the development cycle, for example by using them in assertion statements. Panicking cannot be seen as useful functionality for properly debugged software; a panic is more likely to annoy users than assist them!

Thus, the following is a very bad example of the use of a panic to indicate a problem to a user:

```
_LIT(KTryDifferentMMC, "File was not found in this directory, try
                                            selecting another");
User::Panic(KTryDifferentMMC, KErrNotFound); // Not helpful!
```

The following is a good example of the use of a panic, to highlight a programming error to a developer calling a function in class `Bar` of the `Foo` library, and passing in invalid arguments. The developer can determine which method is called incorrectly and fix the problem:

```
_LIT(KFooDllBarAPI, "Foo.dll, Bar::ConstructL")
User::Panic(KFooDllBarAPI, KErrArgument);
```

Symbian OS itself has a series of well-documented panic categories (for example, `KERN-EXEC`, `E32USER-CBASE`, `ALLOC`, `USER`) and associated error values, the details of which can be found in the Symbian OS Library which accompanies each SDK.

Assertions

Assertions are used to check that assumptions made about code are correct, for example that the states of objects, function parameters or return values are as expected. Typically, an assertion evaluates a statement and, if it is false, halts execution of the code.

On Symbian OS, there is an assertion macro for debug builds only (`__ASSERT_DEBUG`) and another which executes in both debug and release builds (`__ASSERT_ALWAYS`).

The use of assertions in release builds of code should be considered carefully, because assertion statements have a cost in terms of size and speed and, if the assertion fails, will cause code to terminate with a panic, which results in a poor user experience.

The assertion macro tests a statement and, if it evaluates to false, calls the method specified in the second parameter passed to the macro. The method is not hard-coded to be a panic, but it should always terminate the running code and flag up the failure, rather than return an error or leave; panics are the best choice. Assertions help the detection of invalid states or bad program logic so that code can be fixed. It makes sense to stop the code at the point of error, rather than return an error, since it is easier to track down the bug.

This is one example of how to use the debug assertion macro:

```
void CTestClass::EatPies(TInt aCount)
  {
  #ifdef _DEBUG
  _LIT(KMyPanicDescriptor, "CTestClass::EatPies");
  #endif
  __ASSERT_DEBUG((aCount>=0),
        User::Panic(KMyPanicDescriptor, KErrArgument));
  ... // Use aCount
  }
```

It is more common for a class or code module to define a panic function, a panic category string and a set of specific panic enumerators. For example, the following enumeration could be added to `CTestClass`, so as not to pollute the global namespace:

```
enum TTestClassPanic
  {
  EEatPiesInvalidArgument, // Invalid argument passed to EatPies()
  ... // Enum values for assertions in other CTestClass methods
  };
```

A panic function is defined, either as a member of the class or as a static function within the file containing the implementation of the class:

```
static void CTestClass::Panic(TInt aCategory)
  {
  _LIT(KTestClassPanic, "CTestClass");
  User::Panic(KTestClassPanic, aCategory);
  }
```

The assertion in `EatPies()` can then be written as follows:

```
void CTestClass::EatPies(TInt aCount)
  {
  __ASSERT_DEBUG((aCount>=0), Panic(EEatPiesInvalidArgument));
  ... // Use aCount
  }
```

The advantage of using an identifiable panic descriptor and enumerated values for different assertion conditions is traceability. This is particularly useful for calling code using a given library, since the developer may not have access to the library code in its entirety, but merely to the header files. If the panic string is clear and unique, a developer should be able to locate the class which raised the panic and use the panic category enumeration to find the associated failure, which is named and documented to explain clearly why the assertion failed.

Code with side effects should not be called within assertion statements.

```
// Bad use of assertions!
__ASSERT_DEBUG(FunctionReturningTrue(), Panic(EUnexpectedReturnValue));
__ASSERT_DEBUG(++index<=KMaxValue, Panic(EInvalidIndex));
```

The code may well behave as expected in debug mode, but in release builds the assertion statements are removed by the preprocessor, and with them potentially vital steps in the programming logic. Rather than use the abbreviated cases above, statements should be evaluated independently and their returned values then passed into the assertion macros.

Panics, assertions and leaves

Chapter 5 discusses leaves in more detail, but in essence they may legitimately occur under exceptional conditions such as out of memory, insufficient disk space or the absence of a communications link. It is not possible to stop a leave from occurring, so code should implement a graceful recovery strategy and should always catch leaves (using `TRAP` statements).

Programming errors (''bugs'') can be caused by contradictory assumptions, unexpected design errors or genuine implementation errors, such as writing off the end of an array or trying to write to a file before opening it. These are persistent, unrecoverable errors which should be detected and corrected by the programmer rather than handled at run-time. On Symbian OS, the mechanism to do this is to use assertion statements, which terminate the flow of execution of code if an error is detected, using a panic. Panics cannot be caught and handled gracefully.

Exam Essentials

- Know the type of parameters to pass to `User::Panic()` and understand how to make them meaningful

- Understand the use of `__ASSERT_DEBUG` statements to detect programming errors in debug code by breaking the flow of code execution using a panic
- Recognize that `__ASSERT_ALWAYS` should be used more sparingly because it will test statements in released code too and cause code to panic if the assertion fails

References

[Babin 2005 Chapter 7]

Pagonis, John, ''New IPC Mechanisms For Symbian OS'', Symbian Developer Network, ***www.symbian.com/Developer/techlib/papers/newipc/new_ipc_mechanisms_for_symbian_os.pdf***

[Sales 2005 Chapter 1]

Shackman, Mark, ''Publish and Subscribe'', Symbian Developer Network, ***www.symbian.com/developer/techlib/papers/publish&subscribe/PublishAndSubscribe_v1.0.pdf***

[Stichbury 2004 Chapters 10, 13, 15, 16, 20 and 21]

Symbian Developer Library, ''Support for Writable Static Data in Symbian OS'', ***www.symbian.com/developer/techlib/papers/static_data/SupportForWriteableStaticDataInDLLsv1.1.pdf***

11

Client–Server Framework

Introduction

This chapter examines the client–server framework model on Symbian OS. It describes the theory behind the client–server pattern, why it is used on Symbian OS and how it works, the implementation classes involved and the impact, in terms of run-time performance, of using the client–server model.

11.1 The Client–Server Pattern

Critical Information

In the client–server pattern, a client makes use of services provided by a server. The server receives request messages from its clients and handles them, either synchronously or asynchronously.

Servers are typically used to manage shared access to system resources and services. The use of a server is efficient, since it can service multiple client sessions and be accessed concurrently by clients running in separate threads.

A server also protects the integrity of the system, because it can ensure that resources are shared properly between clients and that all clients use those resources correctly. On Symbian OS, when the server runs in its own process it has a separate, isolated address space and the only access a client has to the services in question is through a well-defined interface. By employing a server in a separate process, the operating system can guarantee that badly programmed or malicious clients cannot corrupt any of the resources the server manages. A server must still guard against invalid or out-of-sequence client requests and terminate the offender, typically by raising a panic on the client process.

Most of the system services on Symbian OS, particularly those providing asynchronous functionality, are provided using the client–server framework: for example the window server (for access to UI resources such as the screen and keypad), the serial communications server (for access to the serial ports) and the file system server.

There are several ways in which a server can be started and stopped:

- system servers, for example the file server, are started by Symbian OS as part of OS startup because they are essential to the operating system. They run for the entire time the OS is running and, if they need to terminate unexpectedly, will typically force a reboot of the phone
- application servers, which are only needed when certain applications are running, are started when clients need to connect to them. If an application attempts to start a server that is already running, for example because it has been started by another application, no error results and only a single instance of the server runs. When the server has no outstanding clients, that is, when the last client session closes, it should terminate to save system resources. This type of server is known as a transient server
- Other servers, for example the POSIX server, are required on a per-application basis, and are started with that application and closed when it terminates. Multiple applications may use the same server implementation, but each application will have its own private instance.

The client–server framework can also be used to wrap code which contains writable static data (see Chapter 10).

Exam Essentials

- Know the structure and benefits of the client–server framework:
  - to separate system resources from code which uses them
  - to manage concurrent access and protect the resource from potentially badly programmed or malicious usage
  - to allow access to a resource (such as the file system) by numerous different applications/processes
- Understand the different roles of system and transient servers, and match the appropriate server type to examples of server applications

11.2 Fundamentals of the Symbian OS Client–Server Framework

Critical Information

A Symbian OS server always runs in a separate thread to its clients and often, but not always, runs within a separate process. On Symbian OS, the memory of each process is isolated (see Chapter 10), so when running in different processes, the client and server cannot access each other's virtual address spaces. For this reason, all client–server communication takes place by message passing (to initiate requests) with additional data transfer between the separate threads and/or processes mediated by the kernel.

The communication channel used to pass messages between the client and server is known as a *session*. A session is initiated by a client but the server-side representation is created by the kernel, which also acts as an intermediary for all client–server communication.

The client makes a request to the server, using a message object that identifies the nature of the request and can additionally hold some parameter data. For simple transactions this is sufficient, but for more complex data, inter-thread data transfer functions are used.

A typical server has associated client-side code that formats requests to pass to the server, via the kernel, and hides the implementation details of the private client–server communication protocol. This means that, for example, a ''client'' of the Symbian OS file server (`efile.exe`) is actually a client of the file server's client-side implementation and links against the DLL which provides it (`efsrv.dll`), as shown in Figure 11.1.

Servers are often used to provide asynchronous services to their clients, because they run in a separate thread to their clients. A client may submit a number of asynchronous requests to a server (up to 255, see Section 11.3) but may only ever have one synchronous request outstanding at a time.

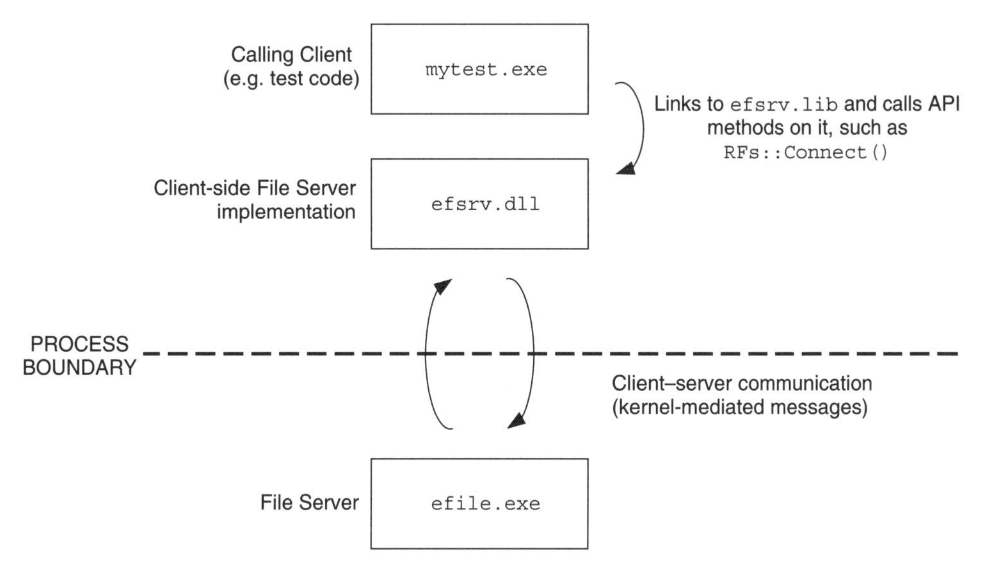

Figure 11.1 File server, client-side implementation and calling client

Exam Essentials

- Know the fundamentals of the Symbian OS client–server implementation:
  - the client and server are always in different threads and often, but not always, in different processes
  - the client–server framework is useful for implementing asynchronous services since the server always runs in a separate thread to the clients
  - the client–server communication channel is known as a session
  - the client makes a series of requests to the server
  - the client and server cannot directly read each other's address space
  - client requests may be synchronous or asynchronous
  - only one synchronous request may be submitted at a time but multiple asynchronous requests may be outstanding.

11.3 Symbian OS Client–Server Classes

Critical Information

This section introduces the classes that are used to implement the Symbian OS client–server framework. The classes discussed are those introduced for secure inter-process communication (IPC) in Symbian OS v8.1b and later.

Client-side server session

The client-side representation of a server session is an R class (see Chapter 4) deriving from `RSessionBase`, the main client-side class. This derives from `RHandleBase`, which is the base class for classes that own handles to other objects, often those created within the kernel. An `RSessionBase` object uniquely identifies a client–server session and is used to send messages to the server.

Most of the methods of `RSessionBase` are protected to ensure that the client-side class that accesses a server does not directly expose access to the server. Instead, the class derives from `RSessionBase` and exports functions that call the protected `RSessionBase` methods. These functions wrap communication with the server and offer more meaningful functionality to their callers. For example, class `RFs`, which derives from `RSessionBase`, provides access to the file server and offers functions such as `RFs::Delete()` or `RFs::GetDir()`, which communicate with the file server without exposing direct access to it.

Class RSessionBase has several overloads of CreateSession() which start a new client–server session. They are typically called by client-side implementation code in an exported method such as Open() or Connect(), which initiates the use of a particular client–server framework.

For example, starting a session with the file server requires a call to RFs::Connect(), which calls RSessionBase::CreateSession(). When the session is opened successfully, corresponding kernel and server-side objects are created.

A server has a unique name which must be passed to RSessionBase::CreateSession() to connect the client to the correct server. The client-side implementation takes care of this, so the calling code does not need to know the name of the server.

In order to stop "spoof" servers from taking the names of critical system servers, the namespace is partitioned into "normal" and "protected" parts. The protected server namespace is defined by all server names beginning with the "!" character. Registering server objects that have names beginning with this character is only permitted for processes possessing the ProtServ capability. That is, the ProtServ capability allows a server process to register with a protected name. (The capability model and platform security are discussed in more detail in Chapter 15.)

Several overloads of RSessionBase::CreateSession() take an integer parameter called aAsyncMessageSlots. This value reserves a number of slots to hold any outstanding asynchronous requests that the client session may have with the server. The maximum number of slots that may be reserved for each server is 255. Other overloads of CreateSession() do not pre-allocate a maximum number of message slots; instead, slots are taken from a kernel-managed global pool, of up to 255 message slots for that server, which is available to the whole system. If the number of outstanding requests to a server exceeds the number of slots in the system pool, or the number reserved for a particular session, the asynchronous request fails to be submitted and completes immediately with the error KErrServerBusy.

There are several overloads of RSessionBase::CreateSession() which take a TSecurityPolicy object. This allows the client code to stipulate criteria for the server to which it will connect, such as the secure ID of the server process or the capabilities of the calling process. The TSecurityPolicy class is discussed in more detail below; secure IDs, capabilities and platform security are discussed further in Chapter 15.

A request to a server is issued through a call to RSessionBase::SendReceive() or RSessionBase::Send(). The SendReceive() method is overloaded to handle synchronous and asynchronous requests. The asynchronous request method takes a TRequestStatus& parameter, while the synchronous version returns the result in a TInt return value. `RSessionBase::Send()` sends a message to the server but does not receive a reply (and in practice, this function is rarely used).

The `Send()` and `SendReceive()` methods take a 32-bit argument to identify the client request, typically defined in an enumeration shared between the client and server code. The request identifiers do not need to be made public, and neither does the order in which parameter data is passed to the server, within a `TIpcArgs` object, as described below. The way in which requests are identified and parameter data is passed to the server constitutes a private *protocol* between the client- and server-side implementations of a particular client–server framework. Callers wishing to use that framework simply use a set of exported public methods, which wrap the protocol and the calls to `RSessionBase::SendReceive()`.

Typically, a class used to access a server has a termination method, which is usually called `Close()`. Internally, this method will call `RHandleBase::Close()`, which sends a disconnection message to the server and sets the session handle to zero. On receipt of this message, the server ends its session with the client by destroying any associated objects that represent the session server-side. If the client has any outstanding requests when `Close()` is called, they are not guaranteed to be completed.

Symbian OS considers requests that are outstanding following a call to `Close()` as a client-side programming error and raises a panic. A server may make an attempt to complete the session's outstanding requests during disconnection; however, this is not always possible and the client should ensure that it cancels all requests before the `Close()` call.

If a client process terminates without calling `RSessionBase::Close()`, the kernel is responsible for any server-side cleanup necessary. Likewise, if a server process terminates unexpectedly, any waiting client requests will be completed with the error code `KErrServerTerminated`, to give the client the opportunity to perform cleanup on any session handles. Once a client or server terminates, even if it later re-starts, sessions used prior to termination cannot be re-used.

If a server supports session sharing, an open client session may be shared by all the threads in a client process and with threads in other processes. However, some servers restrict the session to the thread which connected to the server alone (this was always the case until sharable sessions were introduced in Symbian OS v6.0) and will raise a panic if an attempt is made to share them.

On the client side, if a session can be shared, the first connection to the server should be made as normal using `RSessionBase::CreateSession()`. Once the session is opened, `RSessionBase::ShareAuto()` should be called on it to make it sharable by other threads in the same process. If the session handle is to be passed by IPC to be used by threads within another process (if the server supports globally sharable sessions), `RSessionBase::ShareProtected()` must be called on the handle. Alternatively, a sharable session can simply be created by calling the overload of `RSessionBase::CreateSession()` which takes a `TIpcSessionType` parameter, passing in `EIpcSession_Sharable` or `EIpcSession_GlobalSharable` for a process-sharable or globally sharable session, respectively.

Client-side request arguments

The `TIpcArgs` class is used to package the arguments to be sent to a server. A `TIpcArgs` object constitutes the "payload" for a client–server request; it can package up to four arguments together with information about each argument's type. It is also possible for the object to contain no arguments at all, for requests that have no associated data.

The `TIpcArgs` class has a default constructor and four templated constructors, allowing an object of this type to be constructed for between zero and four arguments. Internally, the arguments are stored in a simple array of four 32-bit values. Consecutive arguments in the constructor's parameter list are put into consecutive slots in the array. The class also has a group of overloaded `Set()` functions which can be used explicitly to set argument values into specific slots within this array. These methods specify the slot to use and the 32-bit value to store, which may be a `TInt`, an `RHandleBase`, a `TAny*`, a `TDes16*` or a `TDes8*`. Section 11.4 will illustrate further how to use class `TIpcArgs`.

Client- and server-side security policy

The `TSecurityPolicy` class represents a generic security policy, and is passed by the client-side implementation code to the server. The owning client would pass one (or more) of these objects to the server to specify which security checks should be done on the client process which is calling the server before allowing it access to the server.

`TSecurityPolicy` can specify a security policy consisting of:

- a check for between 0 and 7 capabilities
- a check for a given Secure Identifier along with 0–3 capabilities
- a check for a given Vendor Identifier along with 0–3 capabilities.

If multiple capabilities are specified, all of them must be present for the security check to succeed.

Server-side client request

The `RMessage2` class represents a client request server-side. The `2` in the name indicates that it is the second version of this class; the original

`RMessage` class was used in versions of Symbian OS up to v8.1b, but has been replaced in later versions, with the introduction of secure inter-process communication.

Class `RMessage2` derives from `RMessagePtr2`, which provides a handle to the message sent by the client. `RMessage2` extends the base class by encapsulating the data associated with the message.

Each client request to the server is represented by a separate object, but clients do not use `RMessage2` objects directly. The `RMessage2` object is the server-side equivalent of the `TIpcArgs` object and request identifier, and is created by the kernel when a client makes a request by, for example, calling `RSessionBase::SendReceive()`, connecting to the server or closing the session.

The 32-bit request identifier (also known as an operation code or *opcode*) can be retrieved by calling `Function()`. The four argument slots that `TIpcArgs` provides may be accessed by calling `Int0()` to return a 32-bit value from the first element of the request array, `Int1()` to return the second element, and so on. In a similar manner, `Ptr0()` returns the contents of the first element in the request array as a `TAny*` pointer, `Ptr1()` the contents of the second element, and so on to the fourth element.

As described above, the layout of the request parameters in the request data array is pre-determined for each client request. For requests where the server expects parameters from the client, it can retrieve the data from the appropriate slot of the `RMessage2` object.

The pointers returned from the methods `Ptr0()`, `Ptr1()`, `Ptr2()` and `Ptr3()` cannot be used directly by the server code if they refer to the address space of a client running in a different process. The server must instead use the overloaded `ReadL()` and `WriteL()` methods of `RMessagePtr2`, which use kernel-mediated inter-process communication to transfer the data.

When the server has serviced a client request, it calls `RMessagePtr2::Complete()` on the `RMessage2` to notify the client. This method wraps a call to `RThread::RequestComplete()` on the client's thread handle. The integer value passed to `RequestComplete()` is written into the client's `TRequestStatus` value and the request semaphore for the client thread is signaled.

Server-side client session

`CSession2` is an abstract base class that represents a client session within the server. For each `RSessionBase`-derived object on the client side, there is an associated `CSession2`-derived object on the server side which is created as part of the client's connect request via `RSessionBase::CreateSession()`. Like `RMessage2`, the class is named to reflect that it is the second version of the class, updated for secure IPC in Symbian OS v8.1.

Classes derived from `CSession2` handle incoming client requests through their implementation of the pure virtual `ServiceL()` method. Typically, this method checks the incoming message parameter to see which request the client has submitted, then handles it by unpacking the message and using the incoming parameters accordingly. When the request has been handled, `Complete()` is called on the associated message object to notify the client thread of request completion.

`CSession2::Server()` provides access to the `CServer2`-derived active object, discussed below.

Server-side active object

The fundamental server-side base class is `CServer2`, which itself derives from `CActive`. It is an abstract base class which must be implemented by each server.

The system ensures that there is only one `CServer2`-derived active object created for each uniquely-named server. `CServer2::StartL()` adds the server to the active scheduler and initiates the first message receive request. The active object accepts requests from all clients of the server as events, receiving notification of each incoming request from the kernel. By calling `Message()` to inspect the `RMessage2` associated with the request, the event-handler method `RunL()` determines whether the `CServer2`-derived object can handle the request itself (for example, for connection or disconnection requests); if not, it directs them to be handled by the `ServiceL()` method of the appropriate server-side `CSession2`-derived class. Having serviced each request, `RunL()` resubmits a "message receive" request and awaits further client requests.

Since `CServer2` is derived from `CActive`, the conventions for active objects should be followed (see Chapter 9). In particular, requests within the server should be handled quickly and efficiently. For example, a long-running asynchronous request should be delegated to another active object running within the server, which processes it as a series of incremental tasks. This will allow the server to remain responsive to incoming requests while processing the long-running request in the background.

Server-side security policy implementation

`CPolicyServer` implements a security policy by extending the normal `CServer2` class. The class checks a received message against a security policy and then performs an action depending on the result of the check. The exact behavior is defined on construction. On receipt of a message, the message opcode number is used to retrieve the associated policy index, which is used as follows, depending on its value:

- `TSpecialCase::EAlwaysPass` – the message is processed as normal by passing it to the `ServiceL()` method of a session (or, in the case of a connection message, by creating a new session)
- `TSpecialCase::ENotSupported` – the message is completed with `KErrNotSupported`
- `TSpecialCase::ECustomCheck` – makes a call to the virtual function `CustomSecurityCheckL()`, which must return a value to indicate whether to process the message or cause it to fail.

If none of the above apply, the policy index is used as a lookup value into an array of security policies given on construction of the `CPolicyServer` object. The associated policy element is used to check the message being processed, and if the message was sent by a client which meets the attributes specified in the policy, the message is processed as normal. Otherwise, `CPolicyServer::CheckFailedL()` is called with the action value specified in the associated policy element. The action will either complete the request with `KErrPermisisonDenied`, panic the client thread or call a custom failure action (`CPolicyServer::CustomFailureActionL()`). For more information about the `CPolicyServer` class, consult the Symbian Developer Library documentation.

Server-side startup

When creating a server, which always runs in separate thread to the client and often in a separate process, it is necessary to allocate both a cleanup stack and an active scheduler for the server. This should be done on startup (see Chapter 9). The cleanup stack is allocated by a call to `CTrapCleanup::New()` (see Section 5.5). The active scheduler is created and installed as described in Section 9.3.

Exam Essentials

- Know the following classes used by the Symbian OS client–server framework, and basic information about the role of each:
  - `RSessionBase`
  - `TIpcArgs`
  - `TSecurityPolicy`
  - `RMessage2`

- `CSession2`
- `CServer2`
- `CPolicyServer`

- Recognize the objects that a server must instantiate when it starts up
- Understand the mechanism used to prevent the spoofing of servers in Symbian OS

11.4 Client–Server Data Transfer

Critical Information

In the rest of this chapter, unless stated otherwise, this discussion assumes that the client and server are running in separate processes, which means that data transfer between them requires inter-process communication (IPC). Under these circumstances, parameter data can never be transferred using simple C++ pointers, because the server never has direct access to the client's address space (or vice versa). Data is passed from the client to the server as a 32-bit value in the request message itself or by passing a pointer to a descriptor in the client address space, which the server accesses using kernel-mediated data transfer.

A set of enumerated values is used to identify which service the client requests from the server, for example:

```
enum THerculeanLabors
  {
  ESlayNemeanLion,
  ESlayHydra,
  ECaptureCeryneianHind,
  ESlayErymanthianBoar,
  ECleanAugeanStables,
  ESlayStymphalianBirds,
  ECaptureCretanBull,
  ECaptureMaresOfDiomedes,
  EObtainGirdleOfHippolyta,
  ECaptureOxenOfGeryon,
  ETakeGoldenApplesOfHesperides,
  ECaptureCerberus,
  ECancelCleanAugeanStables,
  ECancelSlayStymphalianBirds
  };
```

As described in Section 11.3, a `TIpcArgs` object is instantiated and passed to the server with each request. Where parameter data is to be passed to the server, it is passed to the `TIpcArgs` object on construction. Where there are no accompanying request parameters, the `TIpcArgs` object is constructed empty, by default. For example:

```
void RHerculesSession::CancelCleanAugeanStables()
  {
  SendReceive(ECancelCleanAugeanStables, TIpcArgs());
  }
```

The following discussion and example code describe how to implement client-side request code for a range of different parameter types. The full Hercules client–server example code listings can be downloaded from the Symbian Press website for further inspection.

Read-only basic types

The following shows how to pass descriptors and integers to the server

```
// Request which passes a constant descriptor (aDes)
// and a "read-only" integer (aVal) to the server
TInt RHerculesSession::SlayNemeanLion(const TDesC8& aDes, TInt aVal)
  {
  TIpcArgs args(&aDes, aVal);
  return (SendReceive(ESlayNemeanLion,args));
  }
```

Custom types: simple types (T-class objects)

The above methods show how integer and descriptor data are passed to a server, but what about custom data? `RHerculesSession::SlayHydra()` passes an object of type `THydraData`, which is a simple `struct` containing only built-in types, defined as follows:

```
struct THydraData
  {
  TVersion iHydraVersion;
  TInt iHeadCount;
  };
```

`TVersion` is a Symbian OS class defined in `e32cmn.h` as follows:

```
class TVersion
  {
public:
  IMPORT_C TVersion();
  IMPORT_C TVersion(TInt aMajor, TInt aMinor, TInt aBuild);
  IMPORT_C TVersionName Name() const;
public:
  TInt8 iMajor;
  TInt8 iMinor;
  TInt16 iBuild;
  };
```

A `THydraData` object is thus 64 bits in size, which is too large to be passed to the server as one of the 32-bit elements of the request data array. A pointer to the object must instead be *marshaled* across the client–server boundary.

Server-side code should not attempt to access a client-side object directly through a C++ pointer passed from the client to server. When the client and server are running in different processes, the server code runs in a different virtual address space. Under these circumstances, a C++ pointer which is valid in the client process is not valid in the server process; any attempt to use it server-side will result in an access violation. Data transfer between client and server must instead be performed using the inter-thread data transfer methods of class RMessagePtr2.

Before the THydraData object is passed to the server, it must be "descriptorized". Section 7.9 discusses the use of the package pointer template classes TPckg and TPckgC, which can be used to wrap a flat data object such as THydraData with a pointer descriptor, TPtr8. The SlayHydra() method of RHerculesSession creates a TPckg<THydraData> around its THydraData parameter to pass to the server in the request data array. The resulting descriptor has a length equivalent to the size in bytes of the templated object it wraps, and the iPtr data pointer of the TPtr8 addresses the start of the THydraData object.

```
TInt RHerculesSession::SlayHydra(THydraData& aData)
  {
  TPckg<THydraData> data(aData);
  TIpcArgs args(&data);
  return (SendReceive(ESlayHydra, args));
  }
```

Custom types: C- and R-class objects

What if the custom data has variable length, or does not just contain flat data, but owns pointers to other objects, as is common for a C-class object? The following code shows how an object of a C class, containing a pointer to another object or variable-length data, is marshaled from client to server. The CHerculesData class owns two heap descriptor pointers and an integer value; it must have utility code which puts all its member data into a descriptor client-side (*externalization*) and corresponding code to recreate it from the descriptor server-side (*internalization*). (Some of the standard construction code is omitted for clarity in the example, but is included in the code listing available from the Symbian Press website.)

```
class CHerculesData : public CBase
  {
public:
  static CHerculesData* NewLC(const TDesC8& aDes1,
                const TDesC8& aDes2, TInt aVal);
  static CHerculesData* NewLC(const TDesC8& aStreamData);
  ~CHerculesData();
public:
  // Creates an HBufC8 representation of 'this'
```

```
  HBufC8* MarshalDataL() const;
protected:
  // Writes 'this' to the stream
  void ExternalizeL(RWriteStream& aStream) const;
  // Initializes 'this' from the stream
  void InternalizeL(RReadStream& aStream);
protected:
  ... // Constructors omitted for clarity
private:
  HBufC8* iDes1;
  HBufC8* iDes2;
  TInt iVal;
  };

// Creates a CHerculesData initialized with the contents of the
// descriptor parameter. Used server-side
CHerculesData* CHerculesData::NewLC(const TDesC8& aStreamData)
  {
  CHerculesData* data = new(ELeave) CHerculesData();
  CleanupStack::PushL(data);

  // Open a read stream for the descriptor
  RDesReadStream stream(aStreamData);
  CleanupClosePushL(stream);
  data->InternalizeL(stream);
  CleanupStack::PopAndDestroy(&stream); // finished with the stream
  return (data);
  }

CHerculesData::~CHerculesData()
  {
  delete iDes1;
  delete iDes2;
  }

// Creates and returns a heap descriptor which holds
// the contents of 'this'. Used client-side
HBufC8* CHerculesData::MarshalDataL() const
  {
  // Create a dynamic flat buffer to hold this object's member data
  const TInt KExpandSize = 128; // "Granularity" of dynamic buffer
  CBufFlat* buf = CBufFlat::NewL(KExpandSize);
  CleanupStack::PushL(buf);
  RBufWriteStream stream(*buf); // See Chapter 12 for Streams
  CleanupClosePushL(stream);

  ExternalizeL(stream); // Write 'this' to stream
  CleanupStack::PopAndDestroy(&stream);

  // Create a heap descriptor from the buffer
  HBufC8* des = HBufC8::NewL(buf->Size());
  TPtr8 ptr(des->Des());
  buf->Read(0, ptr, buf->Size());

  CleanupStack::PopAndDestroy(buf); // Finished with buf
  return (des); // Transfer ownership to caller
  }
```

```
// Writes 'this' to aStream for marshalling from
// client-side to server-side
void CHerculesData::ExternalizeL(RWriteStream& aStream) const
  {
  if (iDes1) // Write iDes1 to the stream (or a NULL descriptor)
    {
    aStream << *iDes1;
    }
  else
    {
    aStream << KNullDesC8;
    }

  if (iDes2) // Write iDes2 to the stream (or a NULL descriptor)
    {
    aStream << *iDes2;
    }
  else
    {
    aStream << KNullDesC8;
    }

  aStream.WriteInt32L(iVal); // Write iVal to the stream
  }

// For reconstructing CHerculesData from stream
// server-side. Initializes 'this' with the contents of aStream
void CHerculesData::InternalizeL(RReadStream& aStream)
  {
  const TInt KMaxReadSize = 1024; // Limits each to max 1024 bytes
  iDes1 = HBufC8::NewL(aStream, KMaxReadSize); // Read iDes1
  iDes2 = HBufC8::NewL(aStream, KMaxReadSize); // Read iDes2
  iVal = aStream.ReadInt32L(); // Read iVal
  }

// Request passes a C class object
TInt RHerculesSession::SlayErymanthianBoar(const CHerculesData& aData)
  {
  HBufC8* dataDes=NULL;
  TRAPD(r, dataDes = aData.MarshalDataL());
  if (dataDes)
    {
    TPtr8 ptr(dataDes->Des());
    TIpcArgs args(&ptr);
    r = SendReceive(ESlayErymanthianBoar, args);
    delete dataDes;
    }

  return (r);
  }
```

Read–write request parameters

Client-request submission must also differentiate between non-modifiable arguments passing constant data to the server and modifiable arguments used to retrieve data from it.

```
// Request which passes a read/write integer to the server
TInt RHerculesSession::CaptureCeryneianHind(TInt& aCaptureCount)
  {
  TPckg<TInt> countBuf(aCaptureCount);
  TIpcArgs args(&countBuf);
  return (SendReceive(ECaptureCeryneianHind,args));
  }
```

Asynchronous requests

It is important that the client-side data passed to an asynchronous request must not be stack-based. This is because the server may not process the incoming request data until some arbitrary time after the client issued the request. The parameters must remain in existence until that time – so they cannot exist on the stack in case the client-side function which submitted the request returns, destroying the stack frame. This applies not just to read–write parameters, which are typically written towards the end of the request processing, but also to read parameters, since the request may be queued for the server and not even be read until long after the `SendReceive()` call has completed.

The client must also submit a `TRequestStatus` object for notification of request completion.

```
void RHerculesSession::CleanAugeanStables(TRequestStatus& aStatus)
  {
  SendReceive(ECleanAugeanStables, TIpcArgs(), aStatus);
  }
```

Exam Essentials

- Know the basics of how clients and servers transfer data for synchronous and asynchronous requests
- Recognize the correct code to transfer data from a client derived from `RSessionBase` to a Symbian OS server
- Know how to submit both synchronous and asynchronous client–server requests
- Know how to convert basic and custom data types into the appropriate payload which can be passed to the server, as both read-only and read/write request arguments

11.5 Impact of the Client–Server Framework

Critical Information

Session creation overhead

Although a client can have multiple sessions with a server, each client session also consumes resources in both the server and the kernel. For each open client session, the kernel creates and stores an object to represent it, and the server creates an object of its `CSession2`-derived class.

The number of sessions created by a client should be minimized, because each session carries a run-time speed and memory overhead, in the kernel and the client- and server-side threads. Rather than creating and opening multiple sessions on demand, client code should try to share a single session.

For efficiency, where multiple sessions are required, a client–server implementation may provide a *subsession* class to reduce the expense of multiple open sessions. To use a subsession, a client must open a session with the server as normal, and this can then be used to create subsessions which consume fewer resources and can be created more quickly. A typical client subsession implementation derives from `RSubSessionBase` in a similar manner to a client session, which derives from `RSessionBase`. The implementing class provides simple wrapper functions to hide the details of the subsession. A good example of the use of subsessions is `RFile`, which derives from `RSubSessionBase` and is a subsession of an `RFs` client session to the file server. An `RFile` object represents a subsession for access to individual files.

Performance overhead

It is important to be aware of the system performance implications when using the client–server model. The amount of data transferred between the client and server does not cause so much of an overhead as the frequency with which the communication occurs. The main overhead arises because a thread context switch is necessary to pass a message from the client thread to the server thread and back again; if the client and server threads are running in different processes, a process context switch is also involved.

A context switch between threads stores the state of the running thread, overwriting it with the previous state of the replacing thread. If the client and server threads are in the same process, the thread context switch stores the processor registers for the threads. If the client and server are running in two separate processes, in addition to the thread context, the process context (the address space accessible to the thread) must be stored and restored. The MMU must remap the memory chunks for each process, and on some hardware this means that the cache must be flushed. The exact nature of the overhead of a thread or process context switch depends on the hardware in question.

Inter-thread data transfer between threads running in separate processes can also have an overhead because an area of data belonging to the client must be mapped into the server's address space.

Performance improvements

For performance reasons, when transferring data between the client and server, it is preferable where possible to transfer a large amount of data in a single transaction rather than to perform a number of server accesses. There is no upper limit imposed on the amount of data that can be transferred between the client and server, although the saving on context-switching time must be balanced against the memory cost associated with storing and managing large blocks of request data.

For example, Symbian OS components that frequently transfer data to or from the file system generally do not use direct file system access methods such as `RFile::Read()` or `RFile::Write()`. Instead, they tend to use the stream store APIs, described in more detail in Chapter 12. These API methods have been optimized to access the file server efficiently. When storing data to file, they buffer it on the client side and pass it to the file server in one block, rather than passing individual chunks of data as it is received. `RWriteStream` uses a client-side buffer to hold the data it is passed, and only accesses the file server to write it to file when the buffer is full or if the owner of the stream calls `CommitL()`. Likewise, `RReadStream` pre-fills a buffer from the source file when it is created; when the stream owner wishes to access data from the file, the stream uses this buffer to retrieve the portions of data required, rather than calling the file server to access the file.

When writing code which uses a server, such as the file server, it is always worth considering how to make server access most efficient. With the file server, for example, while there are functions to acquire individual directory entries in the file system, it is often more efficient to read an entire set of entries and scan them client-side, rather than call across the process boundary to the file server multiple times to iterate through a set of directory entries. A well-programmed client implementation will present a high-level API so that one transaction performs several actions server-side.

Exam Essentials

- Understand the potential impact on run-time speed from using a client–server session and differentiate between circumstances where it is useful or necessary and where it is inefficient
- Recognize scenarios where an implementation which uses client subsessions with the server would be recommended
- Understand the impact of the context switch required when making a client–server request, and the best way to manage communication between a client and a server to maximize run-time efficiency

References

[Babin 2005 Chapter 9]
[Heath 2006 Chapter 5]
[Stichbury 2004 Chapters 11 and 12]

12

File Server and Streams

Introduction

The file system server, usually known simply as the file server, handles all aspects of managing files and directories on the phone's storage devices, and provides a consistent interface across the ROM, RAM, Flash memory and removable-media devices. The file server runs as a process, `EFILE.EXE`, which receives and executes requests from its callers, who use the client-side implementation classes supplied by `EFSRV.DLL` (for details of the client–server framework, see Chapter 11).

The file server is part of the trusted computer base (TCB) because it also contains the loader, which loads executable files (DLLs and EXEs) from the data-caged `\sys\bin` directory (for details on platform security, see Chapter 15).

12.1 The Symbian OS File System

Critical Information

File server session class

The file server provides the basic services that allow calling code to manipulate drives, directories and files. In order to use the file server, a caller must first create a file server session, represented by an instance of the `RFs` class.

The general pattern for connecting to the file server, using the session to create and use an `RFile` subsession (described in more detail below), and then releasing both, is shown in the following code example.

```
RFs fs;
User::LeaveIfError(fs.Connect()); // Connect the session
```

```
CleanupClosePushL(fs); // Closes fs if a leave occurs

// Create a file
_LIT(KASDExampleIni, "c:\\ASDExample.ini");
RFile file; // A subsession which represents a file, as below
User::LeaveIfError(file.Create(fs, KASDExampleIni,
        EFileRead|EFileWrite|EFileShareExclusive));
CleanupClosePushL(file); // Closes file if a leave occurs

TBuf8<32> buf;
// Submit a read request using the subsession
User::LeaveIfError(file.Read(buf));

// Clean up the RFile subsession and RFs session
// This calls RFile::Close() on file and RFs::Close on fs
CleanupStack::PopAndDestroy(2, &fs); // file, fs
```

The code example uses the cleanup stack, as described in Chapter 5, to ensure that the resources associated with the open file server session and file subsession are leave-safe. If those objects are members of a class, rather than stack-based objects as in the example shown, it is not necessary to use the cleanup stack to protect them because the class destructor will ensure the session and subsession are closed. If a file is not closed explicitly using `RFile::Close()`, it will be closed when the server session associated with it is closed, but it is good practice to clean up any file handle when it is no longer required.

A connected `RFs` session can be used to open any number of files or directories (as subsessions), or to perform any other file-related operations. A file server session can be kept open for the lifetime of an application.

The `RFs` class provides many useful file system-related operations, including:

- `Delete()` and `Rename()` to delete or rename the file specified, or `Replace()` to move it to a different location
- `MkDir()`, `MkDirAll()`, `RmDir()` and `Rename()` to create, remove and rename the directories specified
- `Att()`, `SetAtt()`, `Modified()` and `SetModified()` to read and modify directory and file attributes such as hidden, system or read-only flags
- `NotifyChange()`, an asynchronous request for notification of changes to files, directories or directory entries. `NotifyChange-Cancel()` is used to cancel the outstanding request
- `Drive()`, `SetDriveName()`, `Volume()` and `SetVolumeLabel()` to manipulate drive and volume names
- `ReadFileSection()` to "peek" at file data without opening the file

- `AddFileSystem()`,`MountFileSystem()`,`DismountFileSystem()` and `RemoveFileSystem()` to dynamically add and remove file system plug-ins that extend the file server types Symbian OS can support. Examples of potential file system plug-ins include support for a remote file system over a network, or encryption of file data before it is stored. The plug-in file system modules are implemented as polymorphic DLLs of `targettype fsy` (see Chapter 10).

File handle class

The `RFile` class is a subsession of an `RFs` client session to the file server. An `RFile` object represents access to a named, individual file, providing functions to open, create or replace the file, or to open a temporary file, and to read from and write to it.

- `RFile::Open()` can be used to open an existing file; an error is returned if it does not already exist
- `RFile::Create()` is used to create and open a new file; an error (`KErrAlreadyExists`) is returned if the file already exists
- `RFile::Replace()` creates a file if it does not yet exist, or deletes an existing version of the file and replaces it with a new, empty one if it does
- `RFile::Temp()` opens a new temporary file and assigns a unique name to it.

A common pattern is to call `Open()` to attempt to open an existing file, without replacing any of its data, and to then call `Create()` if it does not yet exist. For example, when using a log file, an existing log file should not be replaced but simply have data appended to it.

```
RFile logFile;
TInt err=logFile.Open(fsSession,fileName,shareMode);
if (err==KErrNotFound) // file does not exist - create it
  err=logFile.Create(fsSession,fileName,shareMode);
```

When opening a file, a bitmask of `TFileMode` values is passed to indicate the mode in which the file is to be used, such as for reading or writing. The share mode indicates whether other `RFile` objects can access the open file, and whether this access is read-only. That is, files may be opened exclusively or shared. For shared files, a region may be locked using `RFile::Lock()` to claim temporary exclusive access to a region of the file, and then unlocked using `RFile::Unlock()`.

If a file is already open for sharing, it can only be opened by another program using the same share mode as the one in which it was originally opened. For example, to open a file as writable, shared with other clients:

```
RFile file;
_LIT(KFileName,"ASDExample.ini");
file.Open(fsSession,KFileName,EFileWrite|EFileShareAny);
```

If another RFile object tries to open ASDExample.ini in EFileShareExclusive or EFileShareReadersOnly mode, access is denied. It can only be accessed in EFileShareAny mode, or through use of the RFs::ReadFileSection() method, which can read from a file without opening it. In fact, RFs::ReadFileSection() can always be used to access a file without opening it, so the contents of a file can never be truly locked, either through use of RFile::Open() methods with EFileShareExclusive flags, or by calling RFile::Lock(). (The RFs::ReadFileSection() method is used, in particular, by Apparc and the recognizer framework to determine the type of a file by rapid inspection of its contents. Recognizers are discussed in Chapter 10.)

The RFile::Write() methods write data from a non-modifiable 8-bit descriptor object (const TDesC8&, see Chapter 7). The RFile::Read() methods read data into an 8-bit descriptor (TDes8&). Both Read() and Write() methods are available in synchronous and asynchronous forms, although neither the asynchronous Read() nor the asynchronous Write() method can be cancelled.

```
// Open ASDExample.ini
_LIT(KASDExample,"c:\\ASDExample.ini");
RFile file;
User::LeaveIfError(file.Open(fs, KASDExample,
           EFileShareExclusive|EFileWrite));

// Write to the file
_LIT8(KWriteData,"Hello ASD");
file.Write(KWriteData);

// Read from the file
 TBuf8<5> readBuf;
 file.Read(readBuf); // readBuf contains "Hello"

 file.Close();
```

There are several variants of RFile::Read() and RFile::Write(). Besides the standard methods shown above, there are overloads which allow the receiving descriptor length to be overridden, the seek position of the first byte to be specified, asynchronous completion, or combinations of these. In all cases, 8-bit descriptors are used. In consequence, RFile is not particularly well suited to reading or writing the rich variety of data types that may be found in a Symbian OS application. This is not an accident, but a deliberate design decision to encourage the use of streams, which provide the necessary functionality and additional optimizations, as described in Section 12.2.

File name manipulation

Files on Symbian OS are identified by a file name specification which may be up to 256 characters in length. As in DOS, a file specification consists of:

- a device, or drive, such as `c:`
- a path, such as `\Document\Unfiled\`, where the directory names are separated by backslashes (\)
- a file name
- an optional file name extension, separated from the file name by a period (`.`).

Symbian OS applications do not normally rely on the extension to determine the file type. Instead, they use one or more UIDs, stored within the file, to ensure that the file type matches the application (see Chapter 10).

Subject to the overall limitation of 256 characters, a directory name, file name or extension may be of any length. The `RFs::IsValidName()` method returns a boolean value to indicate whether a path name is valid.

The Symbian OS file system supports up to 26 drives, from `a:` to `z:`. On Symbian OS phones, the `z:` drive is always reserved for the system ROM and the `c:` drive is always an internal read–write drive, which on some phones may have limited capacity. Drives from `d:` onwards may be internal, or may contain removable media. It may not be possible to write to all such drives; many phones have one or more read-only drives in addition to `z:`, which are used only by the system.

The file system preserves the case of file and directory names, but all operations on those names are case-independent. This means that there cannot be two or more files in the same directory with names which differ only in the case of some of their letters.

File names are constructed and manipulated using the `TParse` class and its member functions. For example, to set an instance of `TParse` to contain the file specification `c:\Documents\Oandx\Oandx.dat`:

```
_LIT(KFileSpec, "c:\\Documents\\Oandx\\Oandx.dat");
TParse fileSpec;
fileSpec.Set(KFileSpec,NULL,NULL);
```

Following this code, the `TParse` getter functions can be used to determine the various components of the file specification. For example:

```
filespec.Drive(); // returns the string "c:"
fileSpec.Path();  // returns the string "\Documents\Oandx\"
```

`TParse::Set()` takes three parameters, the first being the file specification to be parsed. The second and third parameters are pointers to two other `TDesC` descriptors, and either or both may be `NULL`. If present, the file specification pointed to by the second parameter (the *related* file specification) is used to supply any missing components in the first file specification. If used, the third parameter should point to a *default* file specification, from which any components not supplied by the first and second parameters will be taken. Any path, file name or extension may contain the wildcard characters `?` or `*`, respectively representing any single character or any character sequence.

A `TParse` object owns an instance of `TFileName`, which is a `TBuf16<256>` (see Chapter 7); since each character is 2 bytes in size, the data buffer occupies 512 bytes. This is a large object and its use on the stack should be avoided where possible.

Common errors and inefficiencies

A common compile-time error experienced by novice Symbian OS file system users occurs when attempting to use a 16-bit descriptor to read from or write to a file using an `RFile` handle. The `RFile::Read()` and `RFile::Write()` methods take only 8-bit descriptors, which means that wide strings must first be converted. Another common error is the failure to make stack-based `RFs` or `RFile` objects leave-safe, through use of the cleanup stack.

Connections to the file server can take a significant amount of time to set up. Rather than creating multiple sessions on demand, `RFs` session should be passed between functions where possible, or stored and reused. It is also possible to share `RFile` handles within a process, and even between two processes. Allowing an open file handle to be passed from one process to another is a necessary feature in secure versions of Symbian OS.

File system access code can also be made more efficient by remembering the implications of client–server interaction. As described in Chapter 11, efficiency can be improved by minimizing the number of client–server calls. The main technique for achieving this is to transfer more data and thus make fewer file server requests. For example, it is more efficient to read once from a file into one large buffer, and then access and manipulate this client-side, than to make multiple read requests for smaller sections of a file.

In fact, most data-transfer clients of the file server will use the stream store (see Section 12.2), or a relational database, both of which perform this buffering automatically. These components have optimized their use of the file server, and callers that use these APIs rather than access the file server directly gain efficiency automatically.

Exam Essentials

- Understand the role of the file server in the system
- Know the basic functionality offered by class `RFs`
- Recognize code which correctly opens a fileserver session (`RFs`) and a file subsession (`RFile`) and reads from and writes to the file
- Know the characteristics of the four `RFile` API methods which open a file
- Understand how `TParse` can be used to manipulate and query file names

12.2 Streams and Stores

Critical Information

Streams

A Symbian OS stream is the external representation of one or more objects. The process of writing an object's data to a stream is known as *externalization*; the reverse process is termed *internalization*. The stream may reside in a variety of media, including stores (see below), files or memory. In effect, streams provide an abstraction layer over the final persistent storage media.

The external representation of an object's data needs to be agnostic of the particulars of the object's internal storage, such as byte order and data alignment. It is also meaningless to externalize a pointer; it must be replaced in the external representation by the data to which it points. The representation of each item of data must also have an unambiguously defined length. This means that special care is needed when externalizing data types such as `TInt`, whose internal representation may vary in size between different processors and/or C++ compilers.

Storing multiple data items, which may come from more than one object, in a single stream implies that they are placed in the stream in a specific order. Internalization code, which restores the objects by reading from the stream, must therefore follow exactly the same order that was used to externalize them.

The concept of a stream is implemented in the two Symbian OS base classes `RReadStream` and `RWriteStream`, with concrete classes derived from them to support streams that reside in specific media. For example, `RFileWriteStream` and `RFileReadStream` implement a stream that resides in a file, and `RDesWriteStream` and

RDesReadStream implement a memory-resident stream whose memory is identified by a descriptor.

The RReadStream and RWriteStream base classes provide a variety of WriteXxxL() and ReadXxxL() functions to handle specific data types, ranging from 8-bit integers (for example WriteInt8L()) to 64-bit real numbers (for example WriteReal64L()). These functions are called when the << and >> operators are used on the built-in types.

To handle raw data, the stream base classes also provide a range of WriteL() and ReadL() functions, which include overloads to read and write 16-bit Unicode characters rather than bytes. These should be used with caution:

- the raw data is written to the stream exactly as it appears in memory, so it must be in an implementation-agnostic format before calling WriteL()
- a call to ReadL() must read exactly the same amount of data as was written by the corresponding WriteL() call; this can be ensured by writing the length immediately before the data, or terminating the data with a uniquely recognizable delimiter
- there must be a way to acquire the maximum expected length of the data
- the 16-bit WriteL() and ReadL() functions don't provide standard Unicode compression and decompression.

The following example externalizes a TInt16 to a file (named aFileName), which is assumed not to exist before WriteToStreamFileL() is called.

```
void WriteToStreamFileL(RFs& aFs, TDesC& aFileName, TInt16* aInt)
  {
  RFileWriteStream writer;
  writer.PushL(); // put writer on cleanup stack
  User::LeaveIfError(writer.Create(aFs, aFileName, EFileWrite));
  writer << *aInt;
  writer.CommitL();
  writer.Pop();
  writer.Release();
  }
```

Since the only reference to the stream is on the stack, and code following it can leave, it is necessary to push the stream to the cleanup stack, using the stream's (**not** the cleanup stack's) PushL() function. Once the file has been created, the data is externalized using operator <<. After writing, it is necessary to call the write stream's CommitL() function to ensure that any buffered data is written to the stream. Only then is the stream removed from the cleanup stack, using the stream's

Pop() function. Finally, the stream is closed by calling Release(), which frees the resources it has been using.

Operator << is used to externalize the data and operator >> to internalize it. This is a common pattern that can be used for all built-in types except those, like TInt, whose size is unspecified and compiler-dependent. On Symbian OS, a TInt is only specified to be **at least** 32 bits and may be longer, so externalizing it with operator << would produce an external representation of indefinite size. The maximum length of the value should be used to select an appropriate internalization and externalization method. For example, if the value stored in a TInt can never exceed 16 bits, RWriteStream::WriteInt16L() can be used to externalize it and RReadStream::ReadInt16L() to internalize it:

```
TInt i = 1234;
writer.WriteInt16L(i); // writer is RFileWriteStream, initialized
                       // and leave-safe
...
TInt j = reader.ReadInt16L();
... // Cleanup etc
```

Operators << and >> can also be used for any class that provides an implementation of ExternalizeL() and InternalizeL(), which are prototyped as:

```
class TAsdExample
  {
public:
  ...
  void ExternalizeL(RWriteStream& aStream) const;
  void InternalizeL(RReadStream& aStream);
  ...
  }
```

For such a class, externalization can use either:

```
TAsdExample asd;
...
writer << asd; // writer is RFileWriteStream, initialized and
               // leave-safe
```

or:

```
TAsdExample asd;
...
asd.ExternalizeL(writer); // writer is RFileWriteStream,
                          // initialized and leave-safe
```

which are functionally equivalent, and likewise for internalization, using operator >> or InternalizeL().

Operators << and >> can leave, because the resulting operations allocate resources and can thus fail if insufficient memory is available. The operators must be used within a TRAP harness (see Chapter 5) if they are called within a non-leaving function.

Stores

A Symbian OS store is a collection of streams, and is generally used to implement the persistence of objects. The store class hierarchy is illustrated in Figure 12.1, where concrete classes are highlighted in bold text. The abstract base class for all stores is CStreamStore, whose API defines all the functionality needed to create and modify streams. The classes derived from CStreamStore selectively implement the API according to their needs.

Stores can use a variety of different media, including memory (CBufStore), a stream (CEmbeddedStore) and other stores (for example, CSecureStore, which allows an entire store to be encrypted and decrypted). The most commonly used medium is a file.

An important distinction between the different store types is whether or not they are *persistent*. A persistent store can be closed and, at a later time, re-opened and its content accessed. The data in such a store continues after a program has closed it and even after the program itself has terminated. A file-based store is persistent. CBufStore is not persistent, since the store consists of in-memory data which will be lost when it is closed.

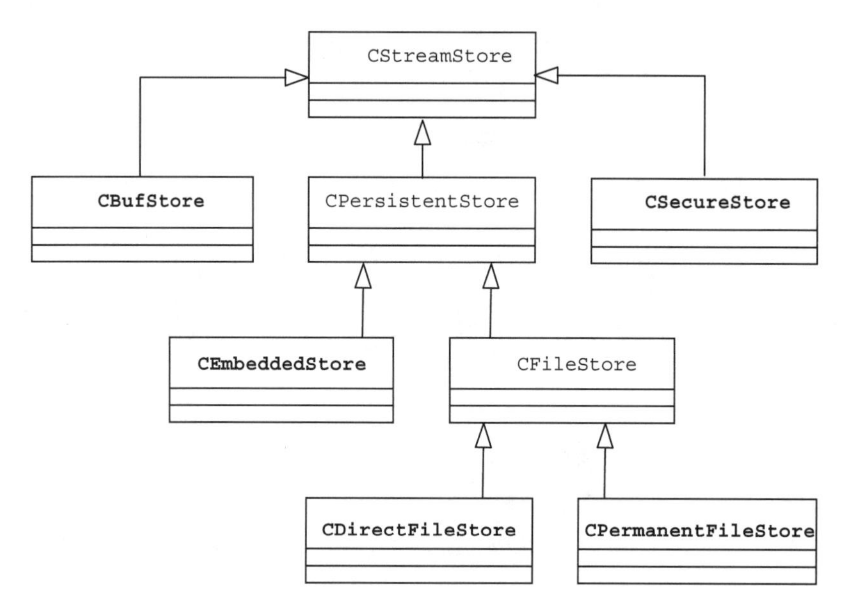

Figure 12.1 Store class hierarchy

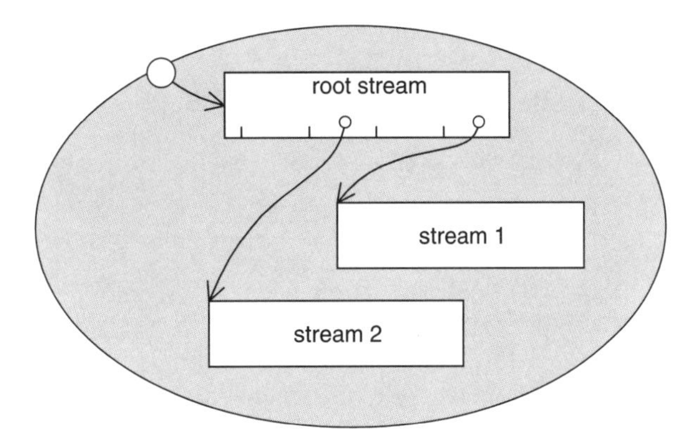

Figure 12.2 Logical view of a persistent store

The persistence of a store is implemented in the `CPersistentStore` abstract class. It defines a *root stream* which is always accessible on opening the store. The root stream contains a *stream dictionary* of pointers to the remaining streams, as illustrated in Figure 12.2. In this way, access to the rest of the data in the store is maintained.

The two file-based stores are `CDirectFileStore` and `CPermanentFileStore`. The main way in which they differ is that `CPermanentFileStore` allows the modification of streams after they have been written to the store, whereas `CDirectFileStore` does not. This difference results in the two stores being used to store persistent data for two different types of application, depending on whether the store or the application itself is considered to contain the primary copy of the application's data.

For an application such as a database, the primary copy of the data is the database file itself. At any one time the application typically holds in memory only a small number of records from the file, in order to display and/or modify them. Any modified data is written back to the file, replacing the original version. Such an application would use an instance of `CPermanentFileStore`, with each record being stored in a separate stream.

Other applications, such as games which store level data, hold all their data in memory and, when necessary, load or save the data in its entirety. Such applications can use a `CDirectFileStore`, since they never modify the store content but simply replace the whole store with an updated version.

Creating a persistent store

The following code illustrates how to create a persistent store. The example creates a direct file store, but creating a permanent file store follows a similar pattern.

```
void CreateDirectFileStoreL(RFs& aFs, TDesC& aFileName, TUid aAppUid)
  {
  CFileStore* store = CDirectFileStore::ReplaceLC(aFs, aFileName,
                                                     EFileWrite);
  store->SetTypeL(TUidType(KDirectFileStoreLayoutUid,
                           KUidAppDllDoc, aAppUid));

  CStreamDictionary* dictionary = CStreamDictionary::NewLC();

  RStoreWriteStream stream;
  TStreamId id = stream.CreateLC(*store);
  TInt16 i = 0x1234;
  stream << i;
  stream.CommitL();
  CleanupStack::PopAndDestroy(); // stream

  dictionary->AssignL(aAppUid,id);

  RStoreWriteStream rootStream;
  TStreamId rootId = rootStream.CreateLC(*store);
  rootStream << *dictionary;
  rootStream.CommitL();
  CleanupStack::PopAndDestroy(2); // rootStream, dictionary

  store->SetRootL(rootId);
  store->CommitL();

  CleanupStack::PopAndDestroy(); // store
  }
```

The example is broken down line-by-line for simplicity. First, the call to `ReplaceLC()` will create the file if it does not exist, otherwise it will replace any existing file. The name of the `ReplaceLC()` method indicates that a reference to the store is left on the cleanup stack, to make it leave-safe (see Chapter 4). In a real application, it might be more convenient to store the pointer to the file-store object in an object's member data, rather than on the stack.

Once created, it is essential to set the store's type:

```
store->SetTypeL(TUidType(KDirectFileStoreLayoutUid,
                         KUidAppDllDoc, aAppUid));
```

The three UIDs in the `TUidType` indicate respectively that the file contains a direct file store, that the store is a document associated with a Symbian OS Unicode application and that it is associated with the particular application whose UID is `aAppUid`. For the file to be recognized as containing a direct file store, it is strictly necessary only to specify the first UID, leaving the other two as `KNullUid`, but including the other two allows an application to be certain that it is opening the correct file.

For comparison, the following code creates a permanent file store:

```
CFileStore* store = CPermanentFileStore::CreateLC(aFs, aFileName,
                                                   EFileWrite);
store->SetTypeL(TUidType(KPermanentFileStoreLayoutUid,
                         KUidAppDllDoc, aAppUid));
```

Note that the `CreateLC()` function is typically used, rather than `ReplaceLC()`, since it is less usual to need to replace a permanent file store.

Creating, writing and closing a stream follow a similar pattern to that discussed above:

```
RStoreWriteStream stream;
TStreamId id = stream.CreateLC(*store);
TInt16 i = 0x1234;
stream << i;
stream.CommitL();
CleanupStack::PopAndDestroy(); // stream
```

The important difference is that an instance of `RStoreWriteStream` (rather than `RFileWriteStream`) must be used to write a stream to a store. The `CreateL()` and `CreateLC()` functions return a `TStreamId`. Once writing the stream is complete, the stream dictionary created earlier in the example can be used to make an association between the stream ID and an externally known UID:

```
dictionary->AssignL(aAppUid,id);
```

Once all the data streams have been written and added to the stream dictionary, the stream dictionary itself must be stored. This is done by creating a stream to contain it, then marking it in the store as the root stream:

```
RStoreWriteStream rootStream;
TStreamId rootId = rootStream.CreateLC(*store);
rootStream << *dictionary;
rootStream.CommitL();
CleanupStack::PopAndDestroy(); // rootStream
...
store->SetRootL(rootId);
```

All that remains is to commit all the changes made to the store and then to free its resources, which in this case is done by the call to the cleanup stack's `PopAndDestroy()`.

```
store->CommitL();
CleanupStack::PopAndDestroy(); // store
```

The store's destructor takes care of closing the file and freeing any other resources.

If a permanent file store is created, it can later be re-opened and new streams added, or existing streams replaced or deleted. To ensure that the modifications are made efficiently, replaced or deleted streams are not physically removed from the store, so the store will increase in size with each such change. To counteract this, the stream store API includes functions to compact the store, by removing replaced or deleted streams.

It is important not to lose a reference to any stream within the store. This is analogous to a memory leak within an application, and results in the presence of a stream that can never be accessed or removed. Arguably, losing access to a stream is more serious than a memory leak, since a persistent file store outlives the application that created it. The stream store API contains a tool, whose central class is `CStoreMap`, to assist with stream cleanup.

Reading a persistent store

The following code opens and reads the direct file store created in the previous example:

```
void ReadDirectFileStoreL(RFs& aFs, TDesC& aFileName, TUid aAppUid)
  {
  CFileStore* store = CDirectFileStore::OpenLC(aFs, aFileName,
                                                         EFileRead);

  CStreamDictionary* dictionary = CStreamDictionary::NewLC();

  RStoreReadStream rootStream;
  rootStream.OpenLC(*store, store->Root());
  rootStream >> *dictionary;
  CleanupStack::PopAndDestroy(); // rootStream

  TStreamId id = dictionary->At(aAppUid);
  CleanupStack::PopAndDestroy(); // dictionary

  RStoreReadStream stream;
  stream.OpenLC(*store, id);
  TInt16 j;
  stream >> j;
  CleanupStack::PopAndDestroy(2); // stream, store
  }
```

After opening the file store for reading, and creating a stream dictionary, the code opens the root stream by calling `RStoreRead-`

`Stream::OpenLC()`, passing in the `TStreamId` associated with root stream, which can be acquired from the store.

Once the root stream is opened, its content can be internalized to the stream dictionary. The dictionary is then used to extract the IDs of the other streams in the store, using the dictionary's `At()` function. Each stream can then be opened individually and internalized as appropriate for the application concerned.

Embedded stores

A store may, in fact, contain an arbitrarily complex network of streams. Any stream may contain another stream – by including its ID – and a stream may itself contain an *embedded* store. This is illustrated in Figure 12.3.

It may be useful to store a collection of streams in an embedded store: from the outside, the embedded store appears as a single stream and can, for example, be copied or deleted as a whole, without the need to consider its internal complexities. An embedded store cannot be modified, and thus behaves like a direct file store – which means that a permanent file store cannot be embedded.

Swizzles

Stores can be used to manage complex data relationships, such as that in a large document which may embed other documents within itself.

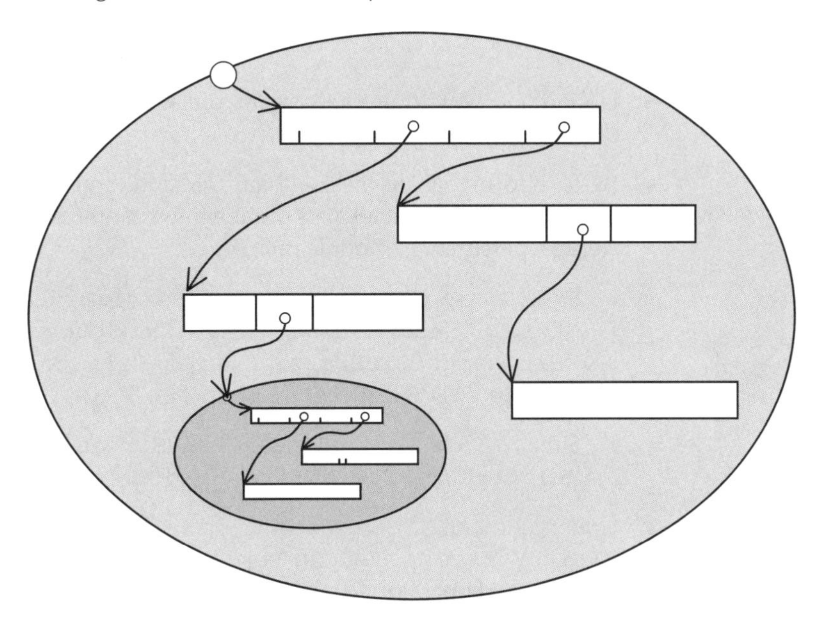

Figure 12.3 An embedded store

An efficient way to manage memory in cases like this is to use a class which maintains a dual representation of the data, and defer loading it into memory from a store until required to do so. The templated swizzle classes `TSwizzleC<class T>` and `TSwizzle<class T>` can be used to represent an object either:

- by stream ID, if the object is not in memory (the stream contains the external representation of that object)
- by pointer, if the object is in memory.

Typically, externalizing a swizzle is a two-stage process which involves:

- externalizing the in-memory object which the swizzle represents, to its own stream
- externalizing the resulting stream ID.

A typical container-type object does not hold a pointer directly to a contained object, but owns a swizzle object which can represent the contained object either as a pointer or as a stream ID.

Exam Essentials

- Know the reasons why use of the stream APIs may be preferred over use of `RFile`
- Understand how to use the stream and store classes to manage large documents most efficiently
- Be able to recognize the Symbian OS store and stream classes and know the basic characteristics of each (for example base class, memory storage, persistence, modification, etc.):
    - Store classes: `CStreamStore`, `CPersistentStore`, `CBufStore`, `CSecureStore`, `CEmbeddedStore`, `CFileStore`, `CPermanentFileStore`, `CDirectFileStore`, `CSwizzle<class T>`, `CSwizzleC<class T>`
    - Stream classes: `RWrite/ReadStream`, `RFileWrite/ReadStream`, `RStoreWrite/ReadStream`
- Understand how to use `ExternalizeL()` and operator << with `RWriteStream` to write an object to a stream, and `InternalizeL()` and operator >> with `RReadStream` to read it back
- Recognize that operators >> and << can leave

References

[Harrison 2004 Chapter 6]
[Sales 2005 Chapter 9]
[Stichbury 2004 Chapter 11]

13
Sockets

Introduction

This chapter gives an initial overview of the function of sockets in communications, and describes the sockets implementation on Symbian OS, including the socket server architecture, the main classes used and the role and features of protocol module plug-ins.

13.1 Introducing Sockets

Sockets were first formally defined by the Lincoln Laboratory MIT, in the 1971 RFC147 paper. The University of California, Berkeley later introduced what would become the de facto sockets API in the BSD Unix 4.2 release during the 1980s.

This section offers a brief introduction to socket terminology, with the focus on network sockets, although sockets may be used for a number of different technologies including infrared, USB and Bluetooth.

Critical Information

Typical socket properties

A socket is a communication endpoint between two or more software processes. Communication is two-way: either a direct *peer-to-peer* relation, where two similar processes communicate; or, more typically, a client–server relationship, where each of the communicating processes performs a different role, for example an internet browser and a web server.

A socket has three characteristics or parameters:

- the communication *domain* is the address family or format of the socket, for example an Internet socket (`KAfInet`) has an address and port number
- the socket *type*, which is typically streaming (connected) or datagram (connectionless); on Symbian OS, these are indicated by use of either `KSockStream` or `KSockDatagram`
- a transport *protocol* that is relevant to the domain; a streaming Internet socket would typically use TCP/IP, indicated on Symbian OS by use of `KProtocolInetTcp`.

Thus a typical TCP/IP Symbian OS socket would be specified as follows:

```
iSocket.Open(iSocketServ, KAfInet, KSockStream, KProtocolInetTcp);
```

The code to open a Symbian OS socket is described in more detail in Section 13.3.

Connection and connectionless

A *connection-oriented* communication protocol establishes an end-to-end connection before any data is sent. The presence of a connection allows certain guarantees because it has a state – for example order of delivery, data arrival and error control. It is often referred to as "reliable".

A *connectionless* communication protocol requires the destination address each time data is sent, and is used for sending datagrams only. As there is no pre-established end-to-end connection, there is no concept of state, and whenever a packet arrives it is treated completely independently of the preceding one. Thus a datagram socket cannot provide guarantees of order, duplication or delivery, and is typically referred to as "unreliable".

Blocking and non-blocking

Sockets may operate in one of two modes, *blocking* and *non-blocking*. An operation on a blocking socket is synchronous, that is the socket does not return control to the calling programming until it has completed the operation. A non-blocking operation is asynchronous and returns control immediately, but requires additional infrastructure to monitor when the data finally does arrive or is fully sent.

The OSI reference model and network sockets

To give a context for how network sockets are used, this section briefly covers the UDP/TCP/IP layers (Transport and Network) of the Open

Systems Interconnection (OSI) reference model. The OSI model is a 7-layer model, encapsulating each layer of communication functionality to provide the layer above it with meaningful semantics and implementation independence:

Layer	Example	Units/Representation
Application	HTTP, SMTP	
Presentation	NCP, SMB	
Session	TLS, RPC	
Transport	TCP, UDP	Segments
Network	IP	Packets
Data Link	Ethernet, 802.11 WiFi	Frames
Physical	100BASE-T, WiFi transceiver	Typically bits

The application, presentation and session layers for TCP/IP are dealt with as a single entity, providing suitable protocols and a relevant API to general applications.

The *transport layer* sends and receives segments of data, while the *network layer* further breaks the segments up into packets on transmitting data, and reassembles packets into segments on receiving data. Each packet contains addressing information and relevant control information as well as the data payload.

The *data link layer* provides synchronization, error- and flow control of frames over the *physical layer*, which deals with the hardware medium.

TCP is a transport-level, **connection-oriented** protocol and provides built-in flow control for reliable transfer of data between network nodes or processes. TCP is packaged and sent over the network-layer IP protocol.

UDP is a transport-level protocol and is **connectionless** and more lightweight than TCP. UDP does not provide any built-in confirmation that the packet has arrived at its destination, nor does it perform any retransmissions on errors, or handshaking of any kind. UDP is used where speed is an issue and reliability is not crucial, for example in network multiplayer games.

IP is the network-level protocol over which both TCP and UDP are layered. TCP and UDP packets reside within the data area of an IP packet. IP is connectionless and data is transferred via packets that flow from a source to a destination.

Symbian OS includes socket support for TCP, UDP and IP for Internet and local network communications, as described in Sections 13.2 and 13.3.

Exceptions and Notes

This section is a brief summary which focuses on network sockets, where typically the protocol is TCP/IP. There are other kinds of socket (for example Bluetooth, Unix, infrared) and additional Address Families which are not generally supported by Symbian OS.

As well as streaming and datagram socket types, there are others including:

- raw sockets, where the header and data payload is provided
- sequence, which ensures the correct order without using a streamed connection
- Reliably Delivered Message (RDM).

Exam Essentials

- Recognize correct high-level statements which define and describe a network socket
- Recognize correct statements about transport independence
- Know the difference between connected and connectionless sockets
- Differentiate between streamed and datagram communication and their relationship with connected/connectionless sockets

13.2 The Symbian OS Sockets Architecture

Critical Information

Symbian OS provides a socket framework which supplies a C++ API similar to the BSD C-based socket API. The socket framework supports communication using the Internet protocol suite and also allows for other types of communication, including Bluetooth, USB and IR. The lower layers of the communications (''comms'') architecture handle any communication differences so that the sockets API can be used in a transport-independent way. This section describes the framework and plug-ins architecture, and discusses the main classes used for sockets programming on Symbian OS.

The sockets framework and protocol module plug-ins

On Symbian OS, the comms system provides support for dynamically loaded protocol modules. The use of sockets means that application code written for use with one protocol can be modified relatively simply to use a different protocol, written for the socket API.

The Symbian OS sockets server (`ESOCK.EXE`) provides communications between addressable endpoints (sockets) and supports a number of different protocols such as Bluetooth (L2CAP and RFCOMM), USB and IR (IrDA, IrTinyTP and IrMUX). The protocols are supplied by plug-in DLLs known as protocol modules (PRTs) which the server loads and unloads as required (for more detail, see Chapter 10). One protocol module can contain multiple protocols: for example, the `TCPIP.PRT` protocol module contains UDP, TCP, ICMP, IP, and DNS, of which UDP and TCP are accessible via sockets to transfer data over IP.

The Symbian OS sockets APIs are published in `es_sock.h`, and calling code uses the client-side implementation classes provided by `esock.dll`. The socket server's client-side APIs make asynchronous calls to the server, which coordinates client access to the socket services and protocol modules that provide support for the particular networking protocol requested. In addition to connecting to sockets, hostname resolution, reading and writing data, the sockets framework also supports querying for protocol information, to allow calling code to determine which sockets protocols are supported on a phone.

`RSocketServ`

`RSocketServ` is the client-side implementation class which communicates with the socket server, analogous to the `RFs` client-side class for file server communication, which is discussed in more details in Chapters 11 and 12.

For code to make requests to the socket server, an object of `RSocketServ` must be instantiated and a session with the socket server established by calling `Connect()`. The prime use for instances of `RSocketServ` is to establish subsession communications for `RSocket` and `RHostResolver`. Any of the `RSocket` objects which are opened using the session are automatically closed when the session is terminated through a call to `RSocketServ::Close()`, and cannot be re-used. However, it is recommended to call `Close()` on each subsession object before closing the session.

`RSocketServ` can be used to pre-load a PRT by calling the asynchronous function `RSocketServ::StartProtocol()`. However, client programs do not normally need to call this function, because loading a protocol is managed automatically by the sockets server when a socket of that protocol is opened. However, some applications (for example those using IrCOMM) may need to ensure that an open socket call will not take a significant amount of time, and so make use of `StartProtocol()`.

`RSocketServ::GetProtocolInfo()` can be used to acquire a comprehensive description of a protocol's capabilities and properties, which is returned in a `TProtocolDesc` object. The number of protocols the socket server is currently aware of can be retrieved by calling `RSocketServ::NumProtocols()`.

`RSocketServ` is not used to send/receive data or establish connections. The `RSocket` subsessions provide APIs to invoke these functions.

RSocket

`RSocket` is the endpoint for all socket-based communications. The class is a subsession of `RSocketServ` and each object instantiated represents a single socket. The methods of `RSocket` correspond, for the most part, with the BSD network API functions.

The client socket interface allows:

- socket opening
- active connecting
- data read from and write to a protocol
- passive connections (through the listen/accept model).

RHostResolver

This class is used for hostname resolution, providing a generic interface to protocol-specific host-resolution services.

For example, IP addresses are hard for people to remember, so ASCII names, known as domain names, are used instead – it's much easier to remember ***www.yahoo.com***, for example, than it is to remember 216.109.118.77! The hostname resolution service used in the internet is the Domain Name System (DNS). DNS translates human-readable domain names to IP addresses. The `RHostResolver` class provides methods for getting an IP address given a domain name, and vice versa. Thus, `RHostResolver::GetByName()` converts the server name to an IP address.

For Bluetooth and infrared, the resolution interface can be used to discover which other devices are in range and available to communicate using those protocols. Queries made through `RHostResolver` objects are packaged in `TNameEntry` descriptors which hold `TNameRecord` objects containing the host name and address.

`RHostResolver` also provides functions to allow getting and setting the hostname of the local device. However, since the interface is generic, and implementation is provided by each protocol module, not all protocol modules provide all services offered by RHostResolver. Functions return KErrNotSupported if the protocol does not support a given operation. RHostResolver is a subsession of an active socket server session, and a host resolver subsession is opened for a specific protocol by passing an appropriate identifier.

Exam Essentials

- Demonstrate a basic understanding of the support for sockets on Symbian OS
- Recognize the characteristics of the RSocketServ, RSocket and RHostResolver classes
- Understand the role and purpose of PRT protocol modules

13.3 Using Symbian OS Sockets

Critical Information

Initialization and socket opening

Before an RSocket subsession can be opened, a session with the socket server must be created through a call to RSocketServ::Connect().

Once a server session is acquired, a socket can be opened by calling one of the overloads of RSocket::Open(), each of which takes the connected socket server session as a parameter. Other parameters that can be specified include the protocol type, the socket type (stream-interface or datagram) and the address family of the socket.

Configuring and connecting sockets

After a socket is open, it is configured differently depending on whether the socket is connectionless or connection-oriented.

Connectionless sockets are configured with a local address and then use special versions of I/O calls that include the remote socket address. The local address is assigned to a connectionless socket using the RSocket::Bind() method. Thus, before it is possible to send or receive data over an unconnected or datagram socket, a call to Open() followed by Bind() must occur. Once the connectionless socket is bound to a local address, it is ready to send or receive data using SendTo() or RecvFrom() respectively, specifying the remote address.

Connected sockets are configured with the address of the remote socket so that they remain tied together for the duration of the connection. A client socket makes a call to the asynchronous RSocket::Connect()

method, passing in the address to connect to the remote socket and waiting for the remote side to complete the connection. If a socket is unbound (that is, if `Bind()` has not yet been called on it) it will automatically have a local address assigned to it.

A server socket must use two sockets. One, which has been bound to a local address by a call to `Bind()`, is used to listen for an incoming connection from a client, using `RSocket::Listen()`. A second, blank socket is then passed to `RSocket::Accept()` to accept a connection once it has been detected. The initial `Listen()` method is synchronous and merely sets up the bound socket with a queue for collecting connection requests. The `Accept()` method is asynchronous and waits for incoming connections. The blank socket is given the handle of the new socket when the asynchronous connection completes, and it may then be used to transfer data.

It is also possible to call `Connect()` on a connectionless socket. If this is done, the `Bind()` call can be omitted and `Send()`, `Write()` and `Recv()` can be used. This allows software initially written to use connection-oriented protocols to be quickly ported to use a connectionless one.

Reading from sockets

Reading and writing with connectionless sockets requires the remote address to be specified in the I/O request. Thus, to read from a connectionless socket, one of the overloads of `RSocket::RecvFrom()` method should be used, rather than `Read()`, `Recv()` or `RecvOneOrMore()` methods of `RSocket`, which should only be used with connected sockets.

`RSocket::RecvOneOrMore()` is an asynchronous request that completes when any data is available from the connection. The receive buffer is specified as an 8-bit descriptor and the received data is added to this buffer. The size of the descriptor is updated to match the number of bytes received.

For stream-interfaced sockets such as TCP, `RSocket::Recv()` will not complete until the entire descriptor (specified by the maximum length of the receive descriptor) is filled with data. This is unlike `RecvOneOrMore()`, which completes when any amount of data is received. So, unless it is known how much data will be received, `Recv()` should not be used for TCP or other stream-interfaced protocols.

`RSocket::RecvFrom()` receives UDP data and supplies not only the data received but also the address of the endpoint that sent the data.

All reading methods are asynchronous and supply a buffer to receive the data. Where there is a corresponding cancellation method, the state of a socket after a read has been cancelled is defined by the characteristics of the protocol.

Writing to sockets

Writing to a connectionless socket is performed using one of the overloads of `RSocket::SendTo()`, which pass the address to which to send the datagram, a buffer of data and flags to control the transfer. Blocking behavior is controlled by passing protocol-specific flags to the call to indicate whether to wait for a receipt.

Writing to connected sockets can use the `RSocket::Write()` method, which does not need to take address information for the remote endpoint. `RSocket::Write()` is a simple asynchronous method which takes a descriptor parameter, passing the entire buffer to the remote socket. As an alternative, the overloaded `RSocket::Send()` methods can also be used for connected sockets. The overloads allow for control over the amount of data sent from the buffer and for passing flags to the underlying protocol module to configure the I/O. All implementations provided for reading from and writing to connected sockets are asynchronous, and each has a corresponding cancellation method.

Closing sockets

When a socket is closed, the protocol layers involved in the socket connection may also shut down. A socket may have pending connection requests or buffers with data ready to be retrieved. To close the socket gracefully, `RSocket::CancelAll()` should be called to cancel any outstanding asynchronous operations waiting on data or connections. The socket can then be closed synchronously by calling `RSocket::Close()` to terminate the socket and cause the operating system to release any resources dedicated to it. This is the typical way to close a connectionless socket and is recommended when there are no pending operations or buffered data waiting.

For connection-oriented protocols, it is usual to disconnect before closing the socket. This can be done by calling `RSocket::Shutdown()`, which is asynchronous and takes a flag to indicate how to close the socket (for example, to drain the socket by waiting for both output and input buffers to empty, or to stop input but drain the output, or vice versa). Although the `Shutdown()` method is asynchronous, socket shutdown cannot be cancelled once it has been called.

Exam Essentials

- Recognize correct patterns for opening and configuring connected and connectionless sockets
- Know which `RSocket` API methods should be used for connected and unconnected sockets to send and receive data

- Know the characteristics of the synchronous and asynchronous methods for closing an `RSocket` subsession

References

[Babin 2005 Chapter 10]
[Harrison 2004 Chapter 8]
Internet Assigned Number Authority, ***www.iana.com***
[Jipping 2002 Chapter 10]
[Stallings 2004 Chapter 4]

14

Tool Chain

Introduction

This chapter gives details of the Symbian OS tool chain and development environment, to reinforce a working knowledge of the subject gained from experience of creating software on the Symbian OS v9.1 platform. Where tool differences exist between Symbian OS v9.1 and previous versions of the platform, the text does not discuss features of the earlier tools, because the ASD exam will be focused on Symbian OS v9.1.

14.1 Build Tools

Critical Information

To build a Symbian OS program, two build files are required:

- the component description file (`bld.inf`)
- the project definition file (`projectname.mmp`).

Build processing

Symbian OS has its own platform-independent build file format (used by `bld.inf`) to specify how a program is built. The `bldmake` tool processes the `bld.inf` component description file to generate a batch file, `abld.bat`, associated with the project definition files specified by `bld.inf`. The `bldmake` tool can be called with the following options:

- `bldmake bldfiles` generates `abld.bat` and associated `.make` files

- bldmake clean removes all files generated by bldmake bldfiles
- bldmake inf displays the basic bld.inf syntax
- bldmake plat displays a list of supported build platforms.

Build platforms represent the various target platforms (and thus binary formats) supported by Symbian OS. When abld.bat is invoked, for example, to build code, the particular platform (emulator or hardware) is specified as an argument. For Symbian OS v9, the most commonly used build platforms are:

- WINSCW, which creates x86-format binaries for running code on the Windows emulator
- GCCE or ARMV5, which create binaries to run on phone hardware, built with the GCCE and RVCT compilers respectively.

When one of the platforms listed is specified as an argument to the abld.bat command, the makefile for that platform is generated and executed.

The syntax of bld.inf is straightforward; its main purpose is to list project definition files and any files that the build tools must export to another location before a build takes place. In the simplest case, bld.inf lists just one project definition file to be built, specified following the PRJ_MMPFILES keyword.

More complex bld.inf files can be made up of a number of sections, using the following keywords:

- PRJ_TESTMMPFILES specifies one or more project definition files for test code (which can be built by invoking abld test build rather than abld build).
- PRJ_EXPORTS lists a series of files to be copied from the project directory to another directory (usually somewhere under \epoc32). The export can be initiated by calling abld export and is performed automatically as part of the abld build command as described later in this section. Likewise, PRJ_TESTEXPORTS lists a series of files to be copied from the project directory to another directory (usually somewhere under \epoc32). The copy can be initiated by calling abld test export and is performed automatically as part of the abld test build command
- PRJ_PLATFORMS can be used to list the platforms that the component supports. If this is not specified, the abld command uses the default set.

Each of the keywords can be specified multiple times in any order. In addition to these keywords, extension makefiles can be used for build tasks which are not offered by the generated makefiles (such as invocation of specialized tools or conversion utilities).

When a change is made to `bld.inf` (for example, to add a new header file under `PRJ_EXPORTS`), `bldmake bldfiles` must be called again to generate a new version of `abld.bat` and any build makefiles that it uses.

The `abld.bat` command can be invoked from the command line with various arguments. The most commonly used are as follows:

- `abld build` combines a number of other arguments (`export`, `makefile`, `library`, `resource`, `target` and `final`) in turn to build the components specified as MMP files under the `PRJ_MMPFILES` specifier in the `bld.inf` file. Likewise, `abld test build` builds those components specified under `PRJ_TESTMMPFILES` in `bld.inf`
- `abld makefile` creates the makefiles for each project specified in `bld.inf`, using the Symbian OS `makmake` tool. The makefiles are then used by `abld` to carry out the various stages of building the component. This command is called each time a component is built, and the makefiles are always re-created, regardless of whether the corresponding `MMP` files have been changed or not since their last creation
- `abld freeze` freezes new DLL exports into `.def` files. For more detail, see Chapter 10 which describes the use of the `IMPORT_C` macro to specify DLL exports and the roles of the module definition file and import library
- `abld clean` erases all the files created by a corresponding `abld target` command (all the intermediate files created during compilation and all the executables and import libraries created by the linker)
- `abld reallyclean` does what `abld clean` does, and also removes files exported by `abld export` and makefiles generated by `abld makefile`, or the `abld test` equivalents.

Each command can be invoked on the projects specified under `PRJ_MMPFILES` in `bld.inf`, by using `abld XXX` (where `XXX` is the command) or on test projects specified under `PRJ_TESTMMPFILES` by using `abld test XXX`.

Project definition files and MMP file syntax

A project definition file, usually referred to as a project's MMP file, is a text file which gives the details needed to build a project. These include the project's source files, the import libraries it needs and the locations of files included through use of #include preprocessor directives. Each statement in a project definition file starts with a keyword. The main keywords will be described for a typical MMP file as follows:

```
TARGET              ASDExample.exe
TARGETTYPE          exe
UID                 0 0xF1101100

CAPABILITY          NONE

SOURCEPATH          ..\src
SOURCE              ASDExampleAppUi.cpp
SOURCE              ASDExampleDocument.cpp
SOURCE              ASDExampleApplication.cpp
SOURCE              ASDExampleView.cpp

SYSTEMINCLUDE       \epoc32\include
USERINCLUDE         ..\inc

SOURCEPATH          ..\data

START RESOURCE      ASDExample.rss
TARGETPATH          \resource\apps
HEADER
END

START RESOURCE      ASDExample_reg.rss
TARGETPATH          \private\10003A3F\apps
END

// Generic Symbian OS libraries
LIBRARY euser.lib efsrv.lib ... // Others omitted for clarity
```

TARGET specifies the name of the file that will be built, ASDExample.exe in the example given above.

TARGETTYPE indicates the type of file to be built: in this case an executable application. The most commonly used Symbian OS target types are DLL, EXE and PLUGIN (ECOM plug-in). Other supported types include PDD and LDD (physical device driver and logical device driver), LIB (a static library whose binary code is included directly in any component that links against it – as compared to a shared library, which is released as a separate binary), and EXEXP (an executable which exports functions).

EPOCEXE and EXEDLL are sometimes used in Symbian OS EKA1 but are no longer necessary in EKA2 because of its improved process emulation.

UID specifies the final two of the target's three Unique Identifiers (UIDs, see Chapter 10) to identify the component. The target will have three UIDs, but the first value (UID1) does not need to be given because it is automatically applied by the build tools according to the TARGETTYPE.

No two executables may have the same UID3 value, and values must be requested from Symbian, which allocates them from a central database. Chapter 15 has more information about the Symbian Signed certification scheme which is used to acquire UIDs.

SECUREID is an optional keyword which is not used in the example above. It is used to define the Secure Identifier (SID) for an executable which, as Chapter 15 explains, is used to identify it. The SID can be specified by a SECUREID statement in the project's MMP file, but if it is not specified the UID3 value is used instead. In MMP files where UID3 is not specified either, KNullUID (=0) is used as both the SID and UID3 value.

VENDORID is an optional keyword which is new in Symbian OS v9.1. It is not used in the example above. An EXE may contain a vendor ID (VID), specified by the VENDORID keyword. The use of a VID identifies the supplier of the binary, but its use is not mandatory.

CAPABILITY is also a new keyword in Symbian OS v9.1. Capabilities are used to restrict the use of certain sensitive system APIs to callers with a particular level of privilege (see Chapter 15 for more information about platform security). The capabilities assigned to an executable are listed following the CAPABILITY keyword in the MMP file. If the CAPABILITY keyword is not used, the capabilities assigned to the binary default to CAPABILITY NONE. The maximum set of capabilities that can be used can be specified using CAPABILITY ALL, but very few components are built with this level of privilege. In general, for code which has a high level of privilege, the maximum set of capabilities specified will be CAPABILITY ALL -TCB.

SOURCEPATH specifies the location of the source or resource files listed using the SOURCE declaration. SOURCEPATH can either be used as a relative location or as a fully qualified path. The keyword can be used multiple times to specify different directories, or can be omitted entirely if all source files are in the same directory as the MMP file.

SYSTEMINCLUDE specifies the directory in which files included in the code using #include <> can be found. On Symbian OS, all global headers should be stored in \epoc32\include or a subdirectory thereof.

USERINCLUDE specifies the directory in which files included in code using #include "" can be found. It can be either a relative path or a fully qualified path. Directories specified with USERINCLUDE are only one of three locations that may be searched for header files, the other directories being that in which the source file which uses the include statement is stored, and the SYSTEMINCLUDE directory.

START RESOURCE...END specifies a resource file, which contains text and specifications of user interface elements, as described later in this section. These keywords replace the use of RESOURCE statements, which were used in MMP files in versions of Symbian OS earlier than v9.1. An application may have several resource files, each of which is specified separately using the START RESOURCE...END block; if the project has a GUI, at least one of these resource files will be needed.

START RESOURCE indicates the beginning of a block of information about an application resource file. The resource file itself should be the same directory as the MMP file or in a directory specified by a preceding SOURCEPATH declaration. The end of the resource file information block is indicated by the END keyword.

In the example shown, the second block specifies a registration resource file for the ASDExample application. This file contains non-localizable information required by the application launcher, such as the application's name, UID and properties. Other information used by the launcher, for example the application's caption (the name displayed for the application in the system shell) and its icons are defined separately so that they can be localized; the location of these definitions is provided in the registration file.

TARGETPATH specifies the build location for a compiled resource (.rsc) as described later in this section. In the example given, the ASDExample.rss resource is compiled to generate output in the \resource\apps directory, which is the standard location for compiled resource files. The second resource, the registration file, is built to \private\10003a3f\apps, which is the standard location for registration information.

The build location for binaries resulting from compilation of C++ code also used to be specified using the TARGETPATH keyword. However, in the secure platform of Symbian OS v9.1 and beyond, all executable code must run from the phone's \sys\bin directory (see Chapter 15). The TARGETPATH keyword is thus now redundant except to build resource files to their appropriate locations.

HEADER is an optional keyword which, when used, causes a resource header file (.rsg) to be created in the system include directory (\epoc32\include) to allow C++ code to use the names of specific resources defined in the associated resource file. In the example given, a header file is generated for access to the resources specified in ASDExample.rss, but no header file is generated for the registration resource file.

LIBRARY lists the import libraries needed by the application. No path needs to be given, and each library statement may contain several libraries, separated by a space. More than one LIBRARY declaration may also be used.

EPOCSTACKSIZE is an optional keyword and is not used in the example above. In previous versions of Symbian OS, the default stack size was 0x5000 bytes; in v9.1, the default value is 0x2000. The EPOCSTACKSIZE keyword can be used, followed by either a decimal or hexadecimal value, to increase the stack size if necessary. However, this option should be used with care, since allocating extra stack space to one application reduces the available space for others. If an application demands a large stack, it should be analyzed for potential improvements and optimizations.

The use of `EPOCSTACKSIZE` only applies when building a binary for the phone and is not supported for WINSCW emulator builds. The stack size is not actually limited on the emulator, since the stack will expand as needed to the much larger limit set by the Windows platform.

`EPOCHEAPSIZE` is an optional keyword and is not used in the example above. It can be used to specify the minimum and maximum sizes (either as decimal or hexadecimal values) of the initial heap for a process; the default sizes are 4 KB minimum and 1 MB maximum. The minimum size specifies the RAM that is initially mapped for the heap's use. The process can then obtain more heap memory on demand until the maximum value is reached. The values specified are rounded up to a multiple of the page size (4 KB).

`EXPORTUNFROZEN` is an optional keyword which can be used by DLLs that are not frozen to have complete `.def` files. The `.lib` import library is created and all exported functions, even unfrozen ones, appear in the import library.

Resource files

On Symbian OS, resource files are typically used to specify the user interface elements of a GUI application, such as the menu bars and dialogs, although they can also be used for any application type. Resource files are also used to define:

- the behavior and functionality of a Symbian OS application
- the application properties that are used by the application launcher
- other literal strings and constant data used in the application, for example dialog text and error messages.

Symbian OS application developers target a wide range of hardware, so to keep platform independence, the resource specifications are kept separate from the executable for each target platform. Resources are specified in a human-readable text file and compiled independently with the Symbian OS resource compiler into a separate binary. This separation reduces the effort required to move applications between different hardware platforms.

The syntax used for resource specification also provides good support for localization by allowing a separation of text from graphics. This not only facilitates translation, but also allows a multi-lingual application to be created without recompilation of the main application code. The application is supplied as a single executable, together with a number of language-specific resource binaries.

Resource files are written as text files and obey a Symbian OS-specific syntax. They are then compiled on their own, using the command-line resource builder tool (`epocrc`, which passes the resource file through the C++ pre-processor and then compiles it with the Symbian OS resource compiler `rcomp`) or as part of the standard build tool chain, from the command-line or within an IDE.

Resource files contain elements which begin with one of the Symbian OS resource keywords such as `RESOURCE`, `STRUCT` or `ENUM`. The resource file is named with the extension `.rss`, and when the resource compiler is invoked on the `.rss` file, it generates two outputs:

- the binary resource file as a `.rsc` file
- a `.rsg` header file in `\epoc32\include`, which is built if the MMP file specifies the `HEADER` keyword in the `START RESOURCE...END` blocks associated with the resource file, as described above. The `.rsg` file contains `#define` statements for each resource defined in the `.rss` file. The header can be used by C++ application code to access elements in the resource binary, by including it using the `#include` pre-processor directive.

A localization file is a text file, typically named with a `.rls` extension (for UIQ) or `.loc` (for S60) and is included directly in the resource file. Other Symbian OS header file types that can be used in a resource file include `.hrh` (a header file that can be shared between C++ and resource files) and `.rh` (a header file used purely by a resource file).

Exam Essentials

- Understand the basic use of `bldmake`, `bld.inf` and `abld.bat`
- Understand the purpose and typical syntax of project definition (MMP) files
- Understand the role of Symbian OS resource and text localization files

14.2 Hardware Builds

Critical Information

The EABI standard

Symbian OS natively runs on ARM processors. Code created for Symbian OS v9 phones must be built with a compiler supporting the Embedded Application Binary Interface (EABI) for the ARM. This is a standard for the interfaces of binary code running in ARM environments, and is intended to allow inter-operation of binaries produced by different compilers that conform to the standard. The standard was designed to give efficient memory usage and data access time, and interoperability between different compiler vendors.

The Symbian OS build tools define native build targets that invoke either:

- a suitable version of the GNU Compiler Collection (GCC), for which the target is identified by the Symbian OS build tools as `GCCE`
- ARM's RealView Compiler Tools RVCT 2.2, for which the target is identified by the Symbian OS build tools as `ARMV5`.

RVCT is intended mostly for Symbian licensees such Nokia and Sony Ericsson to build ROMs for their handset products. GCCE is intended for the majority of Symbian developers; it is delivered with any SDK for phone products based on Symbian OS v9.1 and is also available for free download on the Internet.

The GCCE target compiler

GCCE is a version of the Open Source GNU C++ compiler. The GCCE compiler is intended only for building applications. It cannot be used to compile the full OS.

GCCE can be used either from the command line or invoked from within a development IDE such as CodeWarrior or Carbide.c++. The `GCCE` build target uses the same `.def` file format as the `ARMV5` target. The tools default to looking for `.def` files in the project's `\EABI` directory.

The GCCE compiler is very strict in checking that the source code conforms to the ANSI C++ standard. Some source code which previously compiled with less strict compilers (for example RVCT 2.1) may no longer compile.

The RVCT target compiler

RVCT is used by Symbian to compile Symbian OS and by its licensees to develop ROM-based code. It gives the best performance and smallest code compared to other alternatives, but must be purchased separately. Like GCCE, RVCT can be used either from the command line or invoked from within a development IDE such as CodeWarrior or Carbide.c++.

ARM and THUMB

All current Symbian OS smartphones are based on the ARM processor, which has two instruction sets: a 32-bit set (known as ARM) and a 16-bit set (called THUMB). Code compiled to one set can interoperate with the other. The ARM instruction set is fast but uses more memory per instruction; THUMB is more compact but slower, that is more instructions are required to perform the same work.

The Symbian OS build tools apply the following policy when building projects: kernel-side code is built for ARM, while other code (user-side) is built for THUMB. All code builds into the same ARMV5 subdirectory of `\epoc32\release\`. There are a number of ways to override this policy to build user-side code also for ARM:

- in the `bld.inf` file, the `BUILD_AS_ARM` qualifier can be used to instruct an ARMV5 build not to build the project for the THUMB instruction set, but explicitly for the ARM instruction set

```
PRJ_MMPFILES
ASDExample.mmp BUILD_AS_ARM
```

- to specify that a project should always be built as ARM in an `MMP` files, the keyword `ALWAYS_BUILD_AS_ARM` can be specified.

Exam Essentials

- Understand that the ARM C++ EABI is an industry standard optimized for embedded application development
- Recognize basic information about the RVCT and GCCE compilers, which can be used for target hardware builds
- Understand that ARMV5 supports both 32-bit ARM and 16-bit THUMB instructions, and appreciate the difference with respect to speed and size

14.3 Installing an Application to Phone Hardware

Critical Information

Unlike the Windows emulator, where binaries can simply be copied for testing, the only way to deploy code onto phone hardware on Symbian OS v9 is for the software installer to read it from an installation package or SIS file (`.sis` extension). To create a SIS file, Symbian developers use a package file (`.pkg`) to specify the files and metadata associated with an application to the SIS file creation tool (`MakeSIS`). The package file contains a list of the files, rules, options and dependencies required for the application.

The PC-based `MakeSIS` tool reads the `.pkg` package file and generates a SIS installation file containing all the information necessary for the Symbian OS software installer to install an application to the phone, except for the digital signature. Chapter 15 discusses the role of the Symbian OS software installer in more detail; it is sufficient here to note that most handset manufacturers will require installation packages to be digitally signed before the application contained can be installed onto the phone.

The following is an example of a package file for the `ASDExample` application whose MMP file was described in Section 14.1. In a `.pkg` file, lines preceded by semi-colons are comments and blank lines are ignored.

```
; ASDExample.pkg
; Languages - English and French
&EN, FR

; List of localized vendor names
%{"SymbianPress", "SymbianPress"}

; The non-localized, globally unique vendor name
:"SymbianPress"

; Package header
#{"ASDExample"}, {"ASDExample"}, (0xF1101100), 1, 0, 0, TYPE=SA

; ProductID for UIQ 3.0
(0x101F6300), 3, 0, 0, {"UIQ30ProductID"}

; Files to install for my directory application
; Paths are relative or fully qualified
{"english_info.txt" "french_info.txt"} -
"!:\Documents\ASDExampleGuide.txt"

"..\epoc32\release\gcce\urel\ASDExample.exe"-
"!:\sys\bin\ASDExample.exe"

"..\epoc32\data\Z\Resource\Apps\ASDExample.rsc"-
"!:\Resource\Apps\ASDExample.rsc"

"..\epoc32\data\z\Private\10003a3f\Apps\ASDExample_reg.rsc"-
"!:\private\10003a3f\import\apps\ASDExample_reg.rsc"

IF tkeyboard=1 ; phone has keypad only
"keypad_shortcut_config.txt"-"!:\private\F1101100\shortcut.txt"
ELSEIF tkeyboard=2 ; phone has full QWERTY keyboard
"keyboard_shortcut_config.txt"-"!:\private\F1101100\shortcut.txt"
ELSE ; Display a "No shortcuts are available for this phone" message
    "noshortcut.txt"-"", FILETEXT, TEXTCONTINUE
ENDIF

"readme.txt"-"", FILETEXT, TEXTCONTINUE
```

The line preceded by `&` is the languages section. It lists the supported language variants for the application, using two-character codes as set out in the Languages Table of the Symbian OS Library. In the example above, the `ASDExample.exe` application supports English (`EN`) and French (`FR`).

The sections preceded by % and : are the localized and non-localized vendor names, respectively. Localized vendor names are used in dialogs shown to the user, while the non-localized vendor name is used internally by the software installer.

The line beginning with # is the package header. This line provides the name of the application (localized with one name for each language supported) which is displayed in the installation dialogs, the application's UID (as specified in the MMP file), version information and the installation package type.

The package type indicates the type of installation, since different types have different rules on how files may be installed or uninstalled. The example above uses `TYPE=SISAPP`, which can also be specified using the abbreviation `SA`, or omitted entirely, since it is the default. This type identifies the component to be installed as an application. Other types include a patch type and a partial upgrade type.

The line beginning with a hexadecimal UID in brackets is mandatory and is used to ensure that only applications designed and tested for specific phone hardware can be installed to it. The important values are the hexadecimal UID (`0x101F6300`) and the string in quotes (`UIQ30ProductID`); in the example above, these restrict the installation of `ASDExample.exe` to UIQ 3.0 phones. The equivalent for a phone which runs on the S60 3rd Edition platform is `(0x101F7961), 0, 0, 0, {"Series60ProductID"}`.

Condition blocks (`IF...ELSEIF...ELSE...END`) may be used to control aspects of the installation. In the example above, the condition block tests the `TKeyboard` attribute (from `HalData::TAttribute`) at install time and installs a shortcut configuration file according to whether the phone has a full QWERTY keyboard or a simple numeric keypad.

If the attribute is neither of the values expected, the contents of a text file (`noshortcut.txt`) is displayed to the user during the installation; because it is not installed onto the phone, no destination location is specified for it. The instruction includes some options for the display: `FILETEXT` indicates to display the file during installation (other options include running an executable or creating a blank file in a specified location), and `TEXTCONTINUE` provides a continue button which will dismiss the text file and continue installation (other options include forcing the installation to exit or offering the user the opportunity to abort the installation). This kind of file display can also be useful for showing basic information such as a license agreement at installation time. The example above also uses it, outside a condition statement, to show a `readme.txt` file during the installation of `ASDExample.exe`.

The rest of the package file lists the files to install; the filename before the hyphen indicates a file on the PC, while the location after the hyphen is the destination path on the phone. Specifying an exclamation mark in place of a drive letter in the target filename is recommended, because it means that the user will see a dialog to give a choice of drive on which to install the application. If the drive were to be hard-coded, the user may not have space available on that particular drive, which would make it impossible to install to it, and it is better to offer the user a choice of installation drive.

The line prefixed with `{` specifies a list of files, of which only one will be installed, depending on the language selected by the user during installation.

Exam Essentials

- Recognize the package file format used for creation of SIS installation files

14.4 The Symbian OS Emulator

Critical Information

The Symbian OS emulator is a Windows application called `EPOC.EXE`, which simulates phone hardware on the PC. It is, in effect, a port of the Symbian OS kernel to the Win32 platform. The emulator enables software development for Symbian OS to be substantially PC-based in its early stages, although the final development stages will require the use of phone hardware.

The emulator runs in a single process, which means that on Windows each Symbian OS process is actually loaded as a DLL, and runs inside a separate thread within the single Win32 emulator process, `EPOC.EXE`.

Reasons for using the Symbian OS emulator

The emulator saves time in the early stages of development, because a code development IDE such as CodeWarrior or Carbide.c++ can be used to debug the code and resolve most initial bugs and problems. For example, if a panic occurs in the code, the debugger can provide comprehensive information to diagnose the error condition that caused it.

For hardware testing, an installation file must be created, signed if necessary, transferred to the phone and installed. This can be time-consuming in the early phases of development, when code changes are frequent. The emulator does not need code to be formally installed, which makes the development process much faster.

For emulator builds, the system also writes output to file, which can be inspected for information if a panic occurs, and to check for system warnings such as platform security violations (see Chapter 15). The file is located in the directory associated with the Windows `TEMP` environment variable and is named `epocwind.out` (`%TEMP%\epocwind.out`).

The emulator can be configured through an initialization file called `epoc.ini`, which is stored in `\epoc32\data`, where all configuration files are located for the emulator. For normal use, `epoc.ini` does not need to be modified, but it can be used for example, to add customized virtual drives, change the size of the heap, or map areas of the emulator fascia to act as virtual keys, allowing emulation of phone hardware keys such as the navigation buttons.

Differences between the emulator and Symbian OS phone hardware

The emulator programming environment tends to be more forgiving than that for native code running on phone hardware. For example, code which uses non-constant static variables will compile for the emulator, but will not compile for the ARM platform (see Chapter 10). Some code may run successfully on the emulator but fail on a real phone. For example, it is possible for one process to access the memory of another process on the Windows emulator without causing a memory exception. On the hardware, where memory protection is enforced by Symbian OS memory management, the same code will generate a memory exception.

Symbian has tried to ensure that the Windows emulator provides as faithful an emulation as possible of Symbian OS running on target hardware. The emulator has processes and thread scheduling that are almost identical to those on real Symbian OS phone hardware. However, there are some differences, for example the memory model for a real phone is different to that of the emulator. The underlying hardware is also different, so it is not possible to use the same device driver and hardware abstraction layer code on both the emulator and a real phone. For this reason, the emulator cannot be used for low-level programming such as that for device drivers.

Some other differences in the emulation of Symbian OS on Windows compared to a real hardware platform include the following.

- Bootstrap: on real phones, the first Symbian OS program to run is a bootstrap program, which performs various hardware initialization tasks before starting the kernel. The emulator does not need to perform these tasks, and simply starts the kernel. The remainder of the boot process is similar on the emulator to the native target.
- File system support: the emulator can emulate a range of file system and drive types, but the performance and size of the emulated drives may not be exactly the same as expected for real hardware.
- Floating-point behavior: Symbian OS provides access to IEEE-754 single-precision and double-precision floating-point values through the types `TReal32` (C++ `float` type) and `TReal64` (C++ `double` type) respectively. The emulator is implemented on Intel x86 processors which have floating-point hardware, so this support is used.

Target hardware may or may not have floating-point hardware support; where it does not, the calculations are performed in software. This means that there may be significant performance differences between emulator and hardware versions of the code if it uses floating-point arithmetic.

- Serial ports: the emulator provides emulation of serial ports through Windows serial ports. This is generally adequate for most purposes, but may not provide the same performance as a real device. In particular, some applications have found that high latency times in Windows serial ports have caused some communications data to be dropped.
- Timers: the standard timer resolution is 1/64th second on all platforms, including the emulator. There is also a high-resolution timer, accessed through methods such as `User::AfterHighRes()` and `RTimer::HighRes()`. This has 1 ms resolution on reference hardware, but defaults to 5 ms on the emulator, although it can be changed by setting the timer period (in ms) through the `TimerResolution` variable in the `epoc.ini` configuration file.
- Machine word alignment: the 32-bit RISC architecture used by the phone hardware on which Symbian OS runs requires that 32-bit quantities must be aligned to a 32-bit machine word boundary: in other words, their address must be a multiple of four or an access violation will be generated. This is not the case for code executing on the emulator, which will run successfully.
- Pixel sizes: there is a slight difference in the pixel sizes on the Windows emulator and on phone hardware. This means that text and graphics may be displayed differently on the phone from the way it appears on the emulator.
- USB support: Symbian OS provides USB client support on phone hardware, but there is no such support in the emulator.

Emulator file system

The file system of the phone is mapped to the PC as follows:

- the internal writable drive (`c:`) is usually mapped to `\epoc32\winscw\c`
- the ROM is mapped to `\epoc32\release\winscw\udeb\z` for the debug build, and `\epoc32\release\winscw\udeb\z` for the release build.

The emulator's configuration file can be modified to add other virtual drives if necessary, and the standard drives can be mapped to alternative locations as required.

There is one exception to the file system mapping, which is the location of the executables. On the phone all executables are stored in the `\sys\bin` directory, but on the emulator the executables are loaded from where they are built, that is, the `\epoc32\release\wins\udeb` or `\epoc32\release\wins\urel` directory.

The emulator can also be set up so that it behaves as if a removable media card (for example a Memory Stick or MMC) is present. This can be used to test how an application behaves when reading and writing data to the card, or when the card is removed and/or swapped. It is possible to emulate a user opening and closing the removable media drive door, replacing and removing the media card, and assigning a password to an emulated card.

MMC emulation does not involve access to any kind of hardware interface. Instead, the memory area of each emulated card is represented by a file, a `.bin` type file in the Windows system `temp` directory.

Exam Essentials

- Understand the purpose of the Symbian OS emulator for Windows
- Recognize differences between running code on the emulator and on target hardware

References

[Babin 2005 Chapter 5]
[Stichbury 2004 Chapter 13]
Symbian Developer Library, Symbian OS Tools and Utilities, ***www.symbian.com/developer/techlib/v9.1docs/doc_source/N10356/index.html#etu%2eindex***

15

Platform Security

Introduction

Platform security (often referred to as "PlatSec") was introduced in Symbian OS v9.0 and is fundamental to securing the data and integrity of the phone. The security model operates at the software level to detect and prevent an application from making unauthorized access to hardware, software and system or user data, which might otherwise lock up the phone, compromise confidential user files, or adversely affect other software or the phone network.

Platform security prevents software running on Symbian OS from acting in unacceptable ways, intentionally or unintentionally. It is a system-wide concept which has some impact on every Symbian OS developer, whether they are writing applications, middleware or device drivers.

15.1 The Trust Model

Critical Information

The process is the unit of trust

One of the most fundamental concepts in platform security is the definition of the unit of trust. Symbian OS defines the process as the smallest unit of trust. Processes (most commonly, applications or servers) run on the phone on behalf of the user and may either be built into the phone by the manufacturer or installed on the phone after it has left the factory.

On Symbian OS, platform security controls what a process can do, and restricts it to activities for which it has the appropriate privileges. The operating system will not fulfill a request for a service if a process does not possess the privilege required, because without the appropriate privilege, the process is not deemed trustworthy enough.

The process is deemed the smallest unit of trust because it is the unit of memory protection on Symbian OS. The phone hardware raises a processor fault if access is made in a process to an address not in the virtual address space of that particular process. Symbian OS can trust that a process cannot directly access any other virtual address space, because the hardware prevents it. Thus hardware-assisted protection provides the basis of the software security model. There are, of course, mechanisms for sharing data between processes securely, and these are mediated by the kernel. Inter-process communication is discussed in Chapter 10 and the client–server framework in Chapter 11.

There are four corresponding tiers of trust that apply to processes running on a Symbian OS phone, ranging from completely trustworthy to completely untrustworthy. These are as follows:

- the Trusted Computing Base (TCB)
- the Trusted Computing Environment (TCE)
- other trusted (signed) software
- the rest of the platform (unsigned and therefore untrusted).

Trusted Computing Base (TCB)

The TCB is the most trusted part of Symbian OS, because it controls the lowest level of the security mechanisms and has responsibility for maintaining the integrity of the system. Code within the TCB runs at the highest level of privilege of any on Symbian OS.

To enable verification of the trustworthiness of the TCB, for example through line-by-line code examination, the TCB is kept as small and simple as possible. The Symbian OS TCB has been carefully checked to ensure that the code within it behaves properly and can be considered trustworthy.

The TCB includes the operating system kernel, which looks after the details of each process, including inspecting the set of privileges assigned to it. The file server is also part of the TCB, because it is used to load program code into a process. The privilege information for the code is established by the kernel during this loading process.

For Symbian OS phones which are "closed", that is, which do not support installation of native add-on software, the TCB consists of just the kernel, kernel-side device drivers and file server. However, most Symbian OS phones are "open", and the software installer (SWInstall) also forms part of the TCB. The installer runs when a file is installed from a SIS file package. It extracts files, including program binaries, from the package and has the important role of validating the privileges requested for the program binaries against a digital signature of that package.

Strictly speaking, the memory management unit (MMU) and other security-related hardware features are also part of the TCB. However, these are not discussed further here, because Symbian does not supply the phone hardware.

Most user libraries are not included in the TCB. Only those few which need to be used by the file server or software installer (for example, the cryptography libraries) are given the highest level of trust.

The Trusted Computing Environment (TCE)

The TCE consists of further trusted software provided in the mobile phone by Symbian and other suppliers, such as the UI platform provider and the phone manufacturer. The TCE code is still judged to be trustworthy, but it does not need to run with the highest level of privilege. Thus it is given fewer privileges, and is less trusted, than the code running within the TCB.

Within the TCE, each component has only the privileges needed to carry out a well-defined set of services. By restricting the set in this way, Symbian OS limits the threat posed by any flaw in a server's code or by the possibility that it could become compromised, say by data corruption.

TCE code usually implements these system services in server processes. By requiring selected servers to have certain privileges, it is possible to limit access to sensitive low-level operations to these servers and thereby prevent misuse by other processes.

For example, consider two components of the TCE, the window server and the telephony server:

- the window server (WServ) has privileged access to the screen hardware but has no need to access the phone network
- the telephony server (ETEL) has privileged access to the communications device driver but has no need to access the screen hardware.

The TCE servers provide second-level access to lower-level services by providing APIs to software outside the TCE. For example, a telephony application is not part of the TCE and does not have the same privileges as the telephony server or window server, because it does not communicate directly with the hardware. Instead it uses the APIs each server supplies to perform the appropriate operations.

Symbian OS servers form a cornerstone of the platform security architecture. Each server has a duty to moderate and protect its use of the low-level resources it owns, while also providing mediated access to those resources to clients running in less trusted processes.

Chapter 11 discusses the client–server framework in more detail.

Signed software

It is possible to install software to an "open" Symbian OS phone. Most add-on software lies outside the TCE but still needs certain privileges to use the services the TCE provides.

For example, software which needs access to network services has to open a network socket (see Chapter 13). The socket server (ESOCK) is part of the TCE and handles low-level operations on the network interface. An application that wishes to open a network socket requests the socket server to do so on its behalf. The socket server will first check that the request is from a program which has been granted an appropriate level of trust, because it should not simply grant access to all code wishing to open network sockets – malicious code could otherwise attack the network or other devices. It is important that any application requesting a service be considered trustworthy before the request is granted. A measure of this trust is that the software has been signed.

When an external authority is deciding whether to sign an application to allow it to perform actions (such as opening a network socket) on the phone, the application is not assessed as rigorously as code which runs within the TCB or TCE. That is, the source code itself is usually not inspected line-by-line, although certain functionality of the code is tested and the credentials of the developer are checked. This forms the basis of Symbian Signed, which publishes the tests required on the Symbian Signed Portal (***www.symbiansigned.com***) as described in Section 15.6.

Untrusted software

The trustworthiness of software that is not signed, or is "self-signed" (that is signed, but not by one of the Symbian Trusted Authorities), cannot be determined. This means that it must be considered untrusted by Symbian OS. It does not mean that the software is necessarily malicious or worthless, however. There are many useful operations that can be performed on a phone without calling system services which would require a certain level of privilege.

Symbian OS only makes security checks where necessary to ensure the integrity of the system and protect sensitive services. These checks encompass approximately 40% of all Symbian OS APIs. A solitaire game, for example, does not perform any actions that access sensitive user data or system-critical data or services. Untrusted software can be installed and run on the phone but it is "sandboxed" and cannot perform any actions which require security privileges.

For security, most handset manufacturers will mandate that an application is signed before it can be installed.

The relationship between the TCB, TCE and the rest of the Symbian OS platform is shown in Figure 15.1.

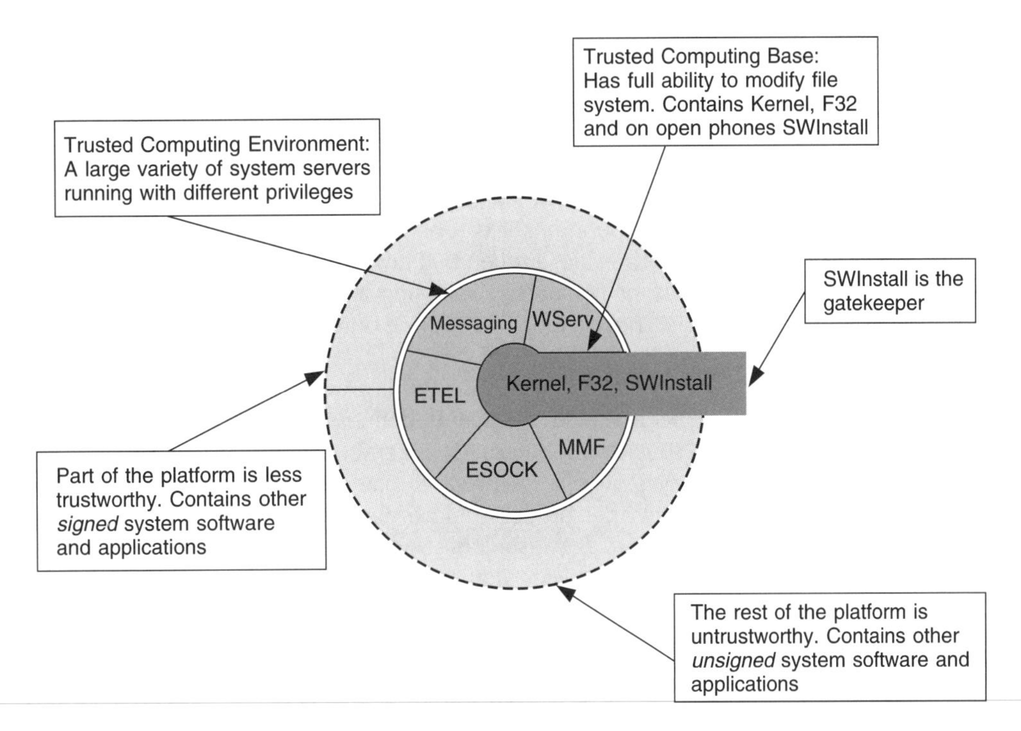

Figure 15.1 Tiers of trust

The figure deliberately does not show the TCB as the center of a set of rings. Although the kernel may normally be thought of as occupying this position, some of its services are available to all processes. Secondly, the file server, like other servers, is both a client of the kernel and available to other processes. The software install component is shown stretching to the outer perimeter, because it acts as the gatekeeper for the phone.

Exam Essentials

- Understand what is meant by the axiom ''a process is a unit of trust'' and how Symbian OS enforces this
- Understand the purpose of the Trusted Computing Base and why it is important
- Recognize that a number of Symbian OS APIs do not require security checks before they can be used
- Know that self-signed software that does not use sensitive system services is ''untrusted'' and can be installed and run on the phone, although it is effectively ''sandboxed''

15.2 Capability Model

Critical Information

As described in Section 15.1, the process is the unit of trust on Symbian OS. Each process is assigned a level of privilege to indicate which security-sensitive operations it can perform. A privilege is designated by a *capability*, which is like a token that gives the holder the right to access a system service or resource. Possessing a capability indicates that the process is trusted not to abuse the service or resource associated with that privilege.

Symbian OS platform security is built around the use of capabilities to represent access privileges, and defines a number of capabilities aligned with specific privileges. The kernel holds a list of capabilities for every running process. A process may ask the kernel to check the capabilities of another process before deciding whether to carry out a service on its behalf. For installable software, the software installer acts as gatekeeper and validates that the program is authorized to use the capabilities it was built with, refusing to install software that does not have the correct authorization (digital signature or, for some capabilities, the user's permission)

Capabilities are *discrete* and *orthogonal*. This means that they do not overlap – they are not a hierarchical set of access tokens with each one adding more privileges until the level of the TCB is reached. Instead, a specific protected resource can be controlled by a single capability. Any process accessing that resource, including processes within the TCB, must have that capability for access to succeed, but the process does not necessarily need any other capability.

Different operations may require different capabilities, even if those operations are implemented using the same APIs. For example, different capabilities are required to access a file depending on its location in the Symbian OS file system (see Section 15.3).

There are three broad categories of capability: TCB capability, system capabilities and user capabilities.

User capabilities

User capabilities are a small group of capabilities deliberately designed to be meaningful to a phone user. These capabilities relate to security concepts that the user can comprehend and make choices about. For example, a user can make decisions about whether to install software with the capability to spend their money by making phone calls, or to access their personal data. A user should not, however, make decisions about capabilities which affect whether the phone works properly, that is, capabilities that affect system services. System-level capabilities are a separate set, described below.

The available user capabilities are summarized in this table:

User Capability	Privilege Granted
`LocalServices`	Access to services over "short-link" connections (such as Bluetooth, USB or infrared). Such services will not normally incur cost for the user
`Location`	Access to data giving the location of the phone
`NetworkServices`	Access to remote services (such as over-the-air data services or WiFi network access). Such services may incur cost for the user
`ReadUserData`	Read access to confidential user data
`UserEnvironment`	Access to live data about the user and their immediate environment
`WriteUserData`	Write access to confidential user data

User capabilities are typically granted to software that makes use of services provided by the TCE. The TCE is responsible for checking and enforcing user capabilities and then performing the requested service.

Platform security is designed so that installable software which needs only user-grantable capabilities does not need to be signed by a trusted authority. The model allows the user instead to grant the capabilities to the software when it is installed (or when it runs and needs to perform an action which requires a particular capability). However, although the user capabilities are designed to be understandable by the user, it may not always be appropriate to offer some choices, depending on the environment in which the phone is being used. The platform security model is deliberately flexible, and mobile phone manufacturers have some discretion in how the user capabilities are configured on their phones, that is, which are actually user-grantable.

System capabilities

The largest group of capabilities is the system capabilities. Granting a system capability allows a process to access sensitive operations, misuse of which could threaten the integrity of the phone. System capabilities are not particularly meaningful to the user and are thus designed to be hidden from them. Installable software which needs system capabilities should be granted those capabilities only after it is submitted to a testing and certification body such as Symbian Signed.

This table summarizes the available system capabilities:

System Capability	Privilege Granted
`AllFiles`	Read access to the entire file system and write access to other processes' private directories (see Section 15.3)
`CommDD`	Direct access to all communications equipment device drivers
`DiskAdmin`	Access to file-system administration operations that affect more than one file or directory (or overall file-system integrity, behavior, etc.)
`Drm`	Access to DRM-protected content
`MultimediaDD`	Access to critical multimedia functions such as direct access to associated device drivers and priority access to multimedia APIs
`NetworkControl`	The ability to modify or access network protocol controls
`PowerMgmt`	The ability to kill any process, to power-off unused peripherals and to cause the phone to go into standby, to wake up, or to power down completely
`ProtServ`	The ability to allow a server process to register with a protected name
`ReadDeviceData`	Read access to confidential network operator, phone manufacturer and device settings
`SurroundingsDD`	Access to logical device drivers that provide input information about the surroundings of the phone

System Capability	Privilege Granted
`SwEvent`	The ability to simulate key presses and pen input, and to capture such events from any program
`TrustedUI`	The ability to create a trusted UI session, and therefore to display dialogs in a secure UI environment
`WriteDeviceData`	Write access to settings that control the behavior of the device

`Tcb` capability

The `Tcb` capability is possessed only by members of the Trusted Computing Base: the kernel, device drivers, file system and, on open Symbian OS phones, the software installer. As described in Section 15.1, the TCB runs with maximum privilege. For this reason, any code running inside the TCB will be assigned all system and user capabilities. Because there are some things that only TCB code can do (such as loading program code), there is one additional capability given only to the TCB. This, the most critical capability, is named `Tcb`.

Assigning capabilities

When building executable code for the Symbian OS secure platform, capabilities can be assigned to it by including them in its MMP (build) file using the `CAPABILITY` keyword. This means that the capabilities are built into the EXE or DLL by the Symbian tool chain. The capabilities can be specified in an inclusive list, for example:

```
CAPABILITY ReadUserData WriteUserData SwEvent
```

Alternatively, a larger set of capabilities can be specified using an exclusive list, for example:

```
CAPABILITY All -Tcb -Drm -DiskAdmin
```

Specifying `CAPABILITY Tcb` is not the same as having all the capabilities, but specifying `CAPABILITY All` does mean possessing capability `Tcb`, as well as all the others.

Once the code is built with the appropriate capabilities, it can then either be included in the ROM of the phone (by the phone manufacturer) or later be installed onto it. It is the handset manufacturers' responsibility to determine that code built into the ROM is of the appropriate level of trust for the capabilities assigned to it, Symbian having performed the necessary security audits for the OS binaries they deliver.

Installable binaries which require system capabilities must be tested and verified by a trusted authority, such as Symbian Signed, and the software installer "gatekeeper" will only allow them be installed if they are accompanied by an appropriate digital signature. Installable binaries which require only user-grantable capabilities, or no capabilities at all, do not in theory need to have a digital signature since they are effectively "sandboxed". However, most handset manufacturers configure Symbian OS platform security to require that the binaries are signed before they can be installed, if only self-signed, to identify the software supplier.

Thus, after a binary is built into ROM or installed to the phone, Symbian OS can assume that it has met the criteria to make it trustworthy for the capabilities it has been assigned. For an EXE, this means that the process will run with the authority to carry out certain privileged operations. For a DLL, this indicates the degree to which it is trusted, and the trustworthiness of the processes into which it can be loaded.

Capability Rule 1

The capabilities of a process never change during its lifetime.

At run-time, the loader, which is part of the TCB, creates a new process by reading the executable code from the file system and determining the set of capabilities required. Once this is done, the capabilities of the process cannot change.

If a process is spawned by another process, it runs independently. It doesn't have the capabilities of the creator process, but has those assigned to it at build time in its MMP file.

Capability Rule 2

A process can only load a DLL if that DLL is trusted with at least the same capabilities as that process

When a process loads a DLL, this does not enlarge or reduce the capability set of the process since, as Rule 1 states, the capabilities of a process never change during its lifetime. A process can load a DLL which has been assigned more capabilities than itself, but the process will not be upgraded to that more trustworthy set; instead, the DLL will be downgraded. A DLL may be loaded into a process which runs with fewer privileges than the DLL was built with, so code within a DLL cannot assume that it will necessarily be running with the capabilities assigned to it. If a DLL requires a particular capability to fulfill its purpose, this should be clearly specified in the documentation for that library.

A process will fail to load a DLL if that DLL does not have at least the same set of capabilities as the process, because if it does not have them, this indicates that the DLL cannot be trusted to run securely within the process. This prevents untrusted, and potentially malicious, code being loaded into sensitive processes (for example, as plug-ins to system servers). A trusted process should not be able to load an untrusted DLL which, for example, uses up all of the available stack or heap (maliciously or simply as the result of poor coding) and prevents the process from functioning.

The DLL loader provides this security mechanism for all processes, relieving them of the burden of identifying which DLLs they can safely load. This is illustrated in Figure 15.2, where the capabilities of the binaries are indicated as "Cn".

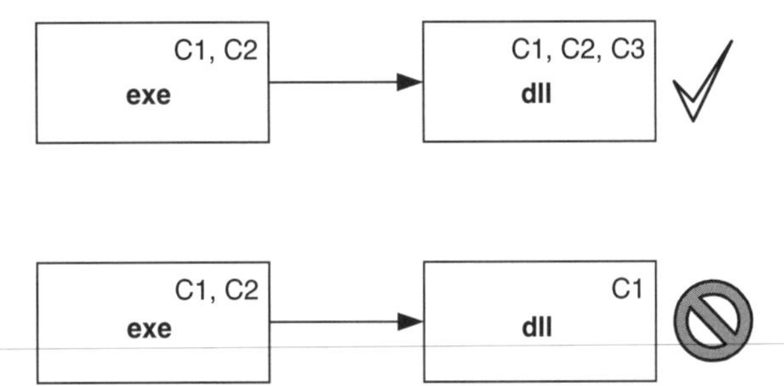

Figure 15.2 Direct loading of DLLs

This rule has some interesting consequences for statically linked DLLs. Most programs using DLLs will be built using static linking, since dynamic loading is primarily used only for a plug-in framework. Static linking resolves references to symbols in the linked DLL at build time, so their run-time loading is more efficient. The most interesting case with regard to capabilities is when one DLL links statically to another. This means that, when the original DLL is loaded into a process, the second DLL is also loaded. Consider the case where the original DLL (DLL1) has a capability that the second DLL (DLL2) does not have, shown in Figure 15.3.

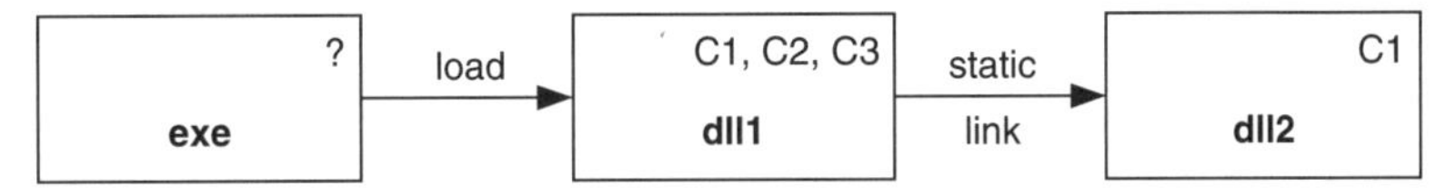

Figure 15.3 Static linking of DLLs

DLL1 can never be loaded into a process which possesses capability C2, because DLL1 links to DLL2, which does not have the trustworthiness to be loaded into an environment with C2. In fact, there is no point declaring capability C2 for DLL1, since it can never be used. On Symbian OS, processes typically reuse loaded DLLs, so this gives rise to an optimization of Rule 2, as follows:

Capability Rule 2A

The loader will only load a DLL that statically links to a second DLL if that second DLL is trusted with at least the same capabilities as the first DLL

This is illustrated by Figure 15.4.

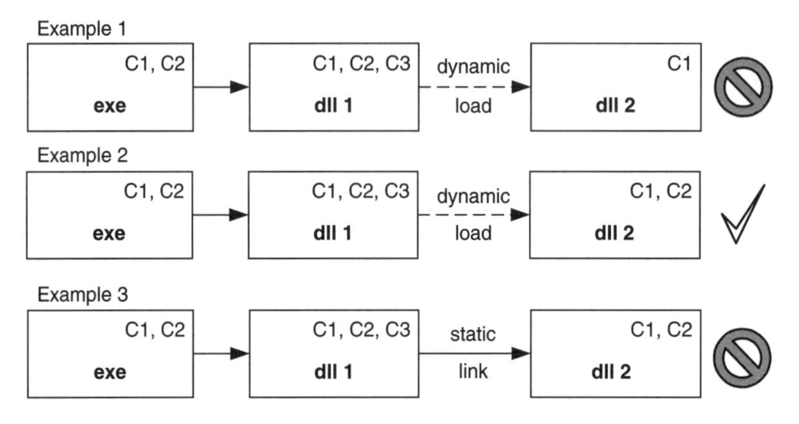

Figure 15.4 Indirect loading of DLLs

Exam Essentials

- Understand the relationship between capabilities and the Trusted Computing Base (TCB)
- Understand the concept of user capabilities and their relationship to the Trusted Computing Environment (TCE)
- Understand the relationship between the TCB/TCE, capability assignment, software install as the ''gatekeeper'' and the role of application signing
- Recognize the different groups of capabilities, demonstrating a broad understanding of the privileges granted
- Recognize how to specify platform security capabilities within an MMP file
- Demonstrate an understanding of the capability rules

15.3 Data Caging

Critical Information

The security of important files is preserved on Symbian OS by file access control known as *data caging*. The caging refers to all types of files, regardless of whether the content is code or data, so Symbian should, perhaps, have called it ''file caging''.

Many system files are critical to the correct functioning and integrity of the phone; others are personal to the user and should remain confidential. Both system and user files can be protected by data caging, which partitions them into secure areas of the file system to prevent corruption or undesirable access.

Data caging is not used to protect all files; there are some public areas of the file system and others which lock away certain files into private areas. The private areas are located under three special top-level paths: `\sys`, `\resource` and `\private`. Access restrictions apply to these directories and to the subdirectories within them, while access to all other paths remains public and no capabilities are required to read from, or write to, them.

The TCB processes have full read–write access to all directories. For other processes, the capabilities required to read or write to particular directories are summarized in this table:

Directories and subdirectories for all drives (`z:`, `c:`, etc.)	**Capability Required to Read**	**Capability Required to Write**
`\sys`	`AllFiles`	`Tcb`
`\resource`	None needed	`Tcb`
`\private\<ownSID>`	None needed	None needed
`\private\<otherSID>`	`AllFiles`	`AllFiles`
`\other` (e.g. `c:\games` or just `c:\`)	None needed	None needed

The `AllFiles` capability allows read access to the entire file system and write access to all subdirectories of `\private`. `Tcb` capability extends this to allow write access to `\sys` and `\resource` and their subdirectories.

The access controls of a file are determined entirely by the location of the directory which contains it, regardless of the drive, and the capabilities of the process attempting to do so. It is not necessary to have explicit access control lists for each file to determine which processes may use it. This also means that, should a developer wish to limit access to a file, it simply needs to be moved to another directory.

Caged file paths: `\sys`

The `\sys\bin` directory is where all program binaries (executables) reside. Executables built into the mobile phone ROM run from z:\sys\bin, while installed software is written into the \sys\bin directory of the c: drive or the equivalent directory on removable media.

To check for tampering of executables installed on removable media, since the binaries could potentially be modified by removing the card from the phone, c:\sys\hash is used to store secure hashes of the binaries.

Only code within the TCB has read/write access to the \sys directory and its subdirectories; code with AllFiles capabilities has read access to \sys. This ensures that only trusted code can access the executables. Binary code stored elsewhere on the phone is not executable, to protect against installation of malware.

One consequence of requiring all executable code to be installed in \sys\bin is that the potential for filename clashes is increased. Developers must ensure their binaries are given unique names, typically by including the SID or UID3, which are unique to a binary (see Section 15.4).

Caged file paths: **\resource**

The \resource directory is for resource files that are strictly read-only, for example bitmaps, fonts and help files. Only the TCB can write into this directory, to provide applications with the assurance that their resource data will not be corrupted, either accidentally or deliberately. No capabilities are required to read from this directory.

As with the \sys directory, files in the \resource directory are either built into the phone ROM or installed onto writable media later by the software installer.

Caged file paths: **\private**

Each process has its own caged file-system area as a subdirectory under \private on each drive. The subdirectory is identified by the SID of the EXE (see Section 15.4). Only processes with the matching SID or those with AllFiles capability possess full read–write access to the directory.

A DLL does not have its own private directory but uses that of the loading process. The Symbian OS file-server access class provides a method to discover the name of the private path, called RFs::PrivatePath().

Exceptions

An installation .pkg file can be used by other parties to specify data to be put into the \private\<SID>\import subdirectory of another application, though only if the import subdirectory already exists on the device.

Exam Essentials

- Understand how data caging works to protect all types of files via the three special directories (`\sys`, `\resource` and `\private`); in particular, that data caging is used to partition all executables in the file system so, once trusted, they are protected from modification
- Understand the implications of data caging for naming executable code
- Recognize that data caging can be used to provide a secure area for an application's data
- Recognize the capabilities needed to read from and write to specific directories and subdirectories
- Know that DLLs do not have a private data-caged area and use that of the process in which they are loaded, and that this directory can be acquired by the DLL using the file system methods `RFs::PrivatePath()`

15.4 Secure Identifier, Vendor Identifier and Unique Identifier

Critical Information

There are several identifiers associated with each executable binary file: SID, VID and some UIDs. Each of these is a 32-bit number.

Unique Identifier (UID)

As discussed in Chapter 10 with respect to DLLs, UIDs are used to uniquely identify a binary file when it executes on Symbian OS. There are three UIDs for any particular executable, which are built into the first 12 bytes of any Symbian OS file to identify it:

- the first UID (UID1) is generated automatically by the Symbian OS build tools, according to the `targettype` of the file stated in the MMP (build) file, for example, `DLL` or `EXE`
- the second UID (UID2) is specified for certain target types, to subdivide them further (for example, for DLLs, there are different UID2 values for polymorphic-interface DLLs and for static-library DLLs)
- the third UID (UID3) is used to identify the binary uniquely. All commercial code will have UIDs assigned through Symbian.

Secure Identifier (SID)

A SID is required to be present and unique for each EXE on the phone, to identify the private directory that the process can access and to uniquely identify the application when it makes an inter-process call.

The SID is similar in intent to the existing UID3 identifier. In fact, the default value of the SID is the UID3 value if the SID is not explicitly specified by use of the `SECUREID` keyword in the MMP file. Typically, the SID is the same as the UID3 value anyway so, to avoid confusion, **it is usually recommended not to specify a SID but simply allow it to default to the UID3 value**.

To keep SID values globally unique, Symbian administers UID3 allocation through a central database which is managed by the Symbian Signed authority. The range of possible 32-bit values has been divided into two ranges – a protected range (`0x00000000–0x7FFFFFFF`) and an unprotected range (`0x80000000–0xFFFFFFFF`).

If a binary file possesses a SID in the protected range, the software installer only permits it to be installed if it has been signed by a certification program, such as Symbian Signed. The certification authority checks that the UIDs built into the binaries submitted for signing do actually belong to the author of the code, and that the author has not previously submitted another application for signing with the same values. It is important to prevent the accidental or deliberate reuse of another application's SID, because clashing SIDs compromise the partitioning of private application data by data caging. Checking for uniqueness of SIDs also prevents one software author spoofing the binaries of another.

Some of the unprotected ID range is not controlled for uniqueness and may be used for unsigned applications and test code. The test range is safe to use for testing because it is not allocated by Symbian Signed, so test code will not clash with legitimate applications. The test range should not be used when shipping self-signed binaries; instead, the allocated range `0xA0000000–0xAFFFFFFF` should be used.

This table shows the UID ranges. The shaded ranges are allocated on request by Symbian Signed (***www.symbiansigned.com***).

UID Range	Intended Use	Status
`0x00000000`	KNullUID	Protected
`0x00000001–0x0FFFFFFF`	Reserved	Protected
`0x10000000–0x1FFFFFFF`	Legacy allocated UIDs, not for V9	Protected
0x20000000–0x2FFFFFFF	UID3/SID range	Protected

UID Range	Intended Use	Status
`0x30000000-0x6FFFFFFF`	Reserved	Protected
`0x70000000-0x7FFFFFFF`	Vendor IDs	Protected
`0x80000000-0x9FFFFFFF`	Reserved	Unprotected
`0xA0000000-0xAFFFFFFF`	UID3/SID range	Unprotected
`0xB0000000-0xE0FFFFFF`	Reserved	Unprotected
`0xE1000000-0xEFFFFFFF`	Development/testing range	Unprotected
`0xF0000000-0xFFFFFFFF`	Legacy UID compatibility range	Unprotected

Vendor Identifier (VID)

The Vendor Identifier (VID) is not required to be unique for every binary. The intention is that multiple executables from a single source could share a VID to identify themselves as originating from a specific software vendor. For example, for Nokia software the VID is `0x101FB657`.

To prevent any vendor from attempting to use the VID of another, a VID may only be used in installable code if it is signed by a certification program, which will verify that the VID specified is appropriate to the author.

The VID of a DLL is not relevant because the VID for a process, and any DLL it loads, is always that of the EXE.

Vendor identifiers can be specified in the MMP file using the `VENDORID` keyword. If none is given, it is assumed to be 0. VIDs cannot be modified after building the application.

Exceptions

SID and VID values are not relevant for DLLs, because they execute within a process (EXE) and use the SID and VID value assigned to that binary instead.

Exam Essentials

- Explain what a Secure Identifier (SID) is, where it is defined and what it is used for
- Understand the similarities and differences between a Secure Identifier (SID), a Vendor Identifier (VID) and a binary's Unique Identifiers (UID)

- Know the rules by which an application is identified, according to the specification of SID, VID and UID
- Understand that SID and VID may be assigned, but are not relevant, to DLLs
- Recognize how to specify VID and SID within an MMP file
- Understand that UIDs are now split into two groups (protected and unprotected ranges) with different implications for test and commercial code

15.5 Application Design for a Secure Platform

Critical Information

A secure application can be defined as follows:

"A secure application is one that its audience can trust without being disappointed. Conversely, an insecure application is one that either its audience does not trust or that disappoints those who trust in it." [Heath 2006 Chapter 4]

Writing a secure application is an important consideration for a developer, because it increases an end-user's confidence in a product. An insecure application may lead to poor distribution and sales, and damage a developer's reputation.

Secure application design has two strands to it:

- analysis of potential threats and their impacts
- analysis and deployment of countermeasures, where appropriate.

Threat analysis

There are several categories of people with an interest in the security of an application:

- The application developer: commercial developers are interested in securing anti-piracy information such as registration codes, because there may be a significant financial impact if this information is compromised. Likewise, artwork and other intellectual property, which may have cost a great deal to create, must be secured, particularly if has been licensed from a supplier under terms which require the content to be protected
- The end-user: the concerns of a user are straightforward, centered on keeping their personal and sensitive data private; keeping that data accessible, by ensuring it is not corrupted or deleted; and preventing

applications from incurring unauthorized financial costs (for example, by sending messages or making calls without the user's consent)

- The owners of other content on the phone, even if the application in question is not expected to interact with their data. No application should be able to access the private data of another application
- The application retailer: retailers will expect an application to protect itself against piracy. They will not want to sell any product which generates end-user complaints about data corruption. Network operators selling applications on their portals will particularly want to avoid distributing applications which give rise to disputed charges for unauthorized calls or messages, or those which disrupt the phone network
- The signing authority: a trusted certification authority, such as Symbian Signed, will require an application to be demonstrably secure before granting a signature to endorse an application as trustworthy.

The following areas must be considered when designing the security of an application:

- Determine which data is not to be disclosed outside the application (for example, registration codes, DRM keys etc.)
- Determine which data is not to be disclosed except to the user or other trusted software, for example the user's contact data or a game's artwork
- Determine which data requires protection against modification, such as a certificate that identifies the bank to which the user is willing to send their account number and password
- Understand which normal behavior could be used in abnormal ways; for example, an image renderer that is trusted to print DRM-protected content should not be able to interact with a module that can print to file, since the combination could be used to obtain an electronic copy of the unprotected data.

Potential attacks can be categorized as follows:

- Attempts to cause inconvenience to the user of the phone, either by preventing expected behavior or causing unexpected behavior such as making expensive calls or causing phone software to crash
- Attempts to trawl the phone for sensitive user or application data
- Attacks that use one application directly, or indirectly, to elevate the attacker's capabilities.

Taking countermeasures

Having performed a threat analysis, a developer will take a range of countermeasures to protect the application from attack. Most countermeasures should be considered early in the application design. For example, the application should typically be modular so that interactions between internal components are predictable and can be controlled, for instance by limiting the capabilities of each module to those required for its particular function.

Early in the design phase, it is necessary to determine which capabilities an application is likely to require since, depending on which are needed, it may be necessary for the application to be certified by a trusted authority such as Symbian Signed. There are several methods used to analyze the capabilities required:

- The general operations the application will perform should be considered and related to the Symbian OS capability set. This is relatively easy because the capabilities are coarse-grained and there are relatively few to choose from. For instance, an instant messaging application may require `NetworkServices` to access the internet and `ReadUserData` to read from the user's address book
- Each API to be used is reviewed to record the capabilities it requires, as described in its documentation
- Trial and error as the code is developed, adding capabilities as they become required. These are determined by using the Windows emulator's platform security settings to write any failed capability checks to the debug log file (see Section 15.6). This approach is generally discouraged as it potentially leads to applications being given more capabilities than they really need, which in turn puts the application at greater risk of attack, weakens the general security model and may cause the application to require unnecessary signing, with cost and time implications.

Most platform-security capabilities have associated security risks, either individually or in combination. An application should be designed so that it cannot be exploited for its capabilities, and should never be assigned more capabilities than it actually needs.

An application should also ensure that its data files are kept private, by using its partitioned `private` directory, which is data-caged from all other processes except those trusted with `Tcb` or `AllFiles` capabilities (see Section 15.3). Applications with import directories (`\private\<SID>\import`), which can have data installed to them by other parties (see Section 15.3, Exceptions), may need their own verification mechanism to decide whether the imported files are trustworthy or not.

Sometimes application data needs to be shared between processes, but it is unusual for it to be shared by many different processes; typically it is only accessible by a well-defined group. In this case, it is sensible to keep the file in a private directory and write a server to provide and police access to the data. A simple way to do this is for a file handle to be passed by the server to any other process that has a sufficient level of trust for it to access the file. Other ways to share data between processes are described in Chapter 10.

Data on removable media can be accessed without going through the Symbian OS protection mechanisms if the media is removed from the mobile phone. Thus, if an application puts data in a private directory on, for example, an MMC, there is nothing to stop the user from removing the card, placing it on a PC and accessing it. Some phones support password-protected cards, but it should not be assumed that all cards will have this protection. Sensitive data should be stored only on internal drives or else should be stored encrypted, for example by designing a secure storage application module, which creates and stores an encryption key on the internal drive and uses this to guarantee that data is secure, regardless of the drive on which it is stored.

Another issue associated with the use of removable media is that the user may modify the contents of the card on a PC and then re-insert it into the phone. This could corrupt the data, so any information that must be tamper-proof must be protected, either by keeping it solely on the internal drive of the phone or by using a cryptographic hash of the data for tamper-detection. Symbian OS uses this technique when program binaries are installed to the `\sys\bin` directory of a removable drive, to ensure that the binaries are not modified when the card is out of the phone. Symbian OS checks the hash of the binary each time it is loaded to ensure it hasn't been tampered with.

It is possible to force binary or data files to be installed to the internal drive by explicitly specifying the drive in the target path of the `.pkg` file, instead of allowing the user to specify the drive at installation time. However, this approach should be used sparingly because it prevents the user from making a choice and making the most of the storage space afforded to them by removable media. It may even prevent them from installing an application at all, if there is insufficient space on the internal drive.

Another route by which data can leave the phone is through a backup to PC, or sometimes by backup to removable media. Backup and restore is an important facility for the user to protect against loss of personal or important data, but it does present a potential security risk, since once data is removed from the phone, it can be viewed and/or tampered with.

For an application, there may be some data which does not need to be backed up, because it can easily be re-created or is transient and likely to change before a restore occurs; for example, the number of hours of voice calls made to date, or a temporary file used for spooling data. Other data simply should not be backed up, for security reasons, such as a PIN for a bank account. It is better to ask the user to re-enter this, if necessary, than to back it up.

Some data can be encrypted before it is backed up, to protect it. If the data must be protected from the user (for example, a DRM key) it must be encrypted with a key which is kept private to the application, stored on the phone and never backed up – otherwise the user could retrieve the key and use it to decrypt the content. Clearly, the encrypted, backed up data can never be restored to any other phone except that which holds the decryption key.

Other data does not need to be protected from the user, but should still be backed up encrypted, to protect the user's data from others who have access to the PC. A good example of this kind of data would be contacts, call records etc. The user can be asked to enter a password to access the backed-up data.

Finally, some application data does not need to be protected in a backup since it is not confidential to either an application or a user. However, in this case, as with all other cases given above, it is important to remember that the data may potentially be tampered with while off the phone. Either the restore operation must check the integrity of data as it is put back on the phone or the application must be prepared to handle unexpected changes to the backed up data which effectively corrupt it.

Exam Essentials

- Demonstrate an understanding of the key considerations when writing a secure application, including the parties interested in application security, typical attacks, countermeasures and secure application design, and the costs of various countermeasures

15.6 Releasing a Secure Application on Symbian OS v9

Critical Information

This section describes some of the steps required, once the binaries are ready, to test a secure application on the Symbian OS v9 emulator and deploy it to phone hardware for further testing and release.

Platform security configuration

The platform-security settings of the Symbian OS emulator are configurable, for example by adding statements to the emulator's initialization file, `\epoc32\data\epoc.ini` (see Chapter 14). The configuration options include:

- PlatSecEnforcement, for enabling/disabling platform security enforcement. When this is enabled, if a platform security check fails, the appropriate action is taken (which is typically to generate a KErrPermissionDenied error or leave). If platform security enforcement is disabled, however, the system continues as if the check had passed
- PlatSecDisabledCaps, for switching off the checking of the capabilities specified after this keyword. The setting instructs the loader to grant the capabilities given in the statement to every process in the system. For example, the statement PlatSecDisabledCaps SwEvent in epoc.ini will disable checking for SwEvent capability
- PlatSecDiagnostics, for enabling/disabling logging of platform-security diagnostic messages to the epocwind.out debug log file when a platform security check fails or generates a warning
- PlatSecEnforceSysBin, to force the emulator to load binaries only from the \sys\bin directory, mimicking the phone hardware.

It can be useful in the early stages of development to suspend platform-security enforcement and enable diagnostic logging of platform-security messages, to determine whether an application has all the capabilities it needs. The default setting for the Windows emulator (and, of course, for phone hardware) is platform security fully enabled and enforced.

SIS file creation: package files, MakeSIS, Makekeys, SignSIS and CreateSIS

Unlike the Windows emulator, where binaries can simply be copied for testing, on phone hardware the only legitimate way to install and test code is for the software installer to read it from an installation package, known as a SIS file.

Symbian OS developers use a Symbian OS package file (.pkg) to specify to the SIS file creation tool (MakeSIS) the files and metadata associated with an application. The package file contains a list of the files, rules, options and dependencies required for the application. The format is documented in the Symbian OS Library and discussed in more detail in Chapter 14.

MakeSIS is a PC-based tool which reads the .pkg package file and generates a SIS installation file containing all the information necessary to install an application to the phone, except the digital signature. Once the SIS file has been created, it must be signed before the contents can be installed onto the phone. It can be either:

- Self-signed, if it requires no capabilities or only capabilities the phone manufacturer has deemed user-grantable (which means that the code is, in effect, untrusted)

- Signed by a testing and certification program such as Symbian Signed, if it requires capabilities which cannot be granted by the user (or if the developer wishes to increase user confidence in the application by having it tested and signed by a certification authority to remove the warning that the SIS file is from an "untrusted source").

To self-sign the SIS file, a developer can use the PC-based Makekeys tool to create a digital certificate and private key. The tool can also be used to generate a PKCS#10 certificate request which can be sent to a trusted authority to be signed. However, most developers taking this option will not use Makekeys to generate a certificate, but will instead apply to Symbian Signed for a *developer certificate*, which will be used as described below for signing installation files for test purposes.

A SIS installation file can be signed using the PC-based SignSIS tool, which takes the SIS file as input along with a certificate and a private key. However, for convenience, Symbian also provides a tool called CreateSIS which is a wrapper around Makekeys, MakeSIS and SignSIS to facilitate the generation and signing of SIS files. The developer may supply a key/certificate pair with which to sign the file, otherwise the tool generates a self-signed certificate and matching key for this purpose.

Developer certificates

The use of trusted authorities to sign SIS installation files and grant capabilities to application code presents an interesting "catch 22" situation. The software installer will only install binary files requiring system capabilities if they are signed by a testing and certification authority, so the developer must submit them to that authority for verification. But in order for the application to pass the exacting certification testing required, it must first have been adequately tested on hardware by the developer. But how can it be tested by the developer if it can't be installed to the phone without the authority's signature?

The solution is to provide an interim certification process which enables the developer to sign a SIS installation file for hardware testing, but limits its installability so that the SIS file cannot be widely distributed. The interim certificates issued are known as *developer certificates* and are issued to a person or organization by Symbian Signed or another trusted authority, such as a handset manufacturer or network operator.

Developer certificates are locked to a specific mobile phone or limited set of phones (on GSM phones, this is done by locking the certificate to a phone's IMEI number). The certificates are limited to a small set of capabilities; typically the developer will have to request special authorization from a handset manufacturer before receiving a developer certificate which allows them to sign binaries for the most sensitive capabilities, `Drm`, `AllFiles`, `CommDD`, `DiskAdmin`, `MultimediaDD`,

Network Control and of course Tcb. Developer certificates are also valid for only a limited period, currently six months, from the date of issue. These restrictions prevent a developer from using the certificates to distribute software commercially.

Symbian Signed

An application which requires capabilities that cannot be granted by the user cannot be installed to a phone protected by platform security unless it has been validated by a trusted certification authority, such as Symbian Signed. Symbian Signed was launched in March 2004 and has seen significant industry endorsement.

The signing process uses a standardized Public Key Infrastructure (PKI) to provide easy to use, robust security and authentication. Two certificates are used for developer authentication and application authorization.

The first certificate used is the Authenticated Content Signing (ACS) Publisher ID, which has a chain of trust to a public root certificate. This is used to authenticate the identity of the developer, who will be asked to provide information about him- or herself, and the contact details of referees who will verify the information supplied.

The second certificate used is the Content ID, which has a chain of trust to a root certificate built onto the ROM of Symbian OS v9 mobile phones. This is used to provide application authorization by granting capabilities to the application based on verification by Symbian Signed. The application will only be authorized as trustworthy of the capabilities it requires after it has been verified by an independent test house against a set of industry-agreed criteria.

The Symbian Signed test criteria can be downloaded from the Symbian Signed Portal so that a developer can first test the application to ensure that it complies with the requirements before submitting it to the test house. The tests comprise two main elements. The first is generic testing, which is performed on all submitted applications to validate their stability and adherence to required processes (such as correct use of UIDs, efficient memory use and uninstall cleanup). The second element consists of tests targeted for the specific capabilities requested for the application. The more capabilities an application requires, the more targeted testing it will require, which will mean a more costly test cycle, thus providing another reason why a developer should limit the capabilities requested to the minimum set that an application actually needs.

To summarize, the Symbian Signed process consists of the following steps:

- Development and internal testing using developer certificates to sign code before installing to hardware
- Developer authentication using the ACS Publisher ID

- Symbian Signed Testing against industry-agreed criteria
- Signing with the Content ID certificate to confirm trust

Exceptions

Developer certificates and Symbian Signed testing and certification are not required to install or release binaries that require no capabilities, or only capabilities from the user-grantable set. The developer can both test and release their untrusted "sandboxed" applications (for example freeware which uses no capabilities) simply by self-signing them. Some developers may still submit their binaries to Symbian Signed for testing and certification, which means their code is deemed "trusted" and installs without warnings, to increase a user's confidence in the application.

Exam Essentials

- Understand the basic process of testing and releasing a signed Symbian OS v9 application

15.7 The Native Software Installer

Critical Information

The software installer is the "pinch point" for getting an add-on ("after-market") software package correctly installed on to a Symbian OS v9 phone. The software installer engine is provided by Symbian; a user-interface layer is added by the mobile phone manufacturer, so the controls look quite different on, say, an S60 phone and a UIQ phone. The installer is often referred to as the *native software installer* to indicate that it installs software which runs directly on Symbian OS, rather than on higher layers (as, for example, Java MIDlets running within a Java Virtual Machine).

The software installer's key responsibilities are:

- To validate and install native software packages (known as SIS files, as described in Section 15.6)
- To validate software delivered in a pre-installed form on media cards
- To handle upgrades and removals, and provide package management services to the rest of the platform.

The installer performs a series of checks before installation to verify that the SIS file has not been tampered with and that the application binaries' capabilities do not exceed those allowed, based on the signature which accompanies the SIS package. If the check passes, the installer places the files in the correct data-caged directories on the phone, as described in Section 15.3. This is why the installer is known as the "gatekeeper" and forms part of the TCB, as described in Section 15.1.

The introduction of platform security to Symbian OS v9 added significant new requirements to the software installer and the SIS installation packages it works with. Symbian took the opportunity to redesign the basic structure of the SIS files, and changed the internal format from previous releases. This means that pre-v9 SIS files are not compatible with the Symbian OS v9 software installer. This may appear to be a big compatibility break, but the binaries in an old SIS file are not compatible with those required to run on Symbian OS v9 anyway, because of the binary break the new version of the OS introduces. The v9 software installer checks the binaries and aborts installation if it encounters an old-style (pre-v9) incompatible file or one built for the wrong target (for example, a binary built to run on the emulator in a package designed to install code to hardware, or vice versa).

Exam Essentials

- Recognize the key functions of the v9 Native Software Installer, including the compatibility break in SIS file format between v9 and previous versions of Symbian OS

References

[Heath 2006 Chapters 2, 3, 4, 5, 8 and 9]
[Stichbury 2004 Chapter 13]
Symbian Signed Portal, ***www.symbiansigned.com***

16

Compatibility

Introduction

This chapter examines compatibility at source level and at binary level. Fundamentally, any interface is a contract that has to be maintained with its users; breaking the contract breaks compatibility. A good Symbian developer should understand what can, and what cannot, be modified in an interface to extend a component without breaking either source or binary compatibility.

This chapter is a collection of essential programming rules to ensure that code can be future-proof and behave as a good "compatibility citizen".

16.1 Levels of Compatibility

Critical Information

Forward and backward compatibility

Compatibility works in two directions, forwards and backwards. When a component is updated in such a way that other code that used the original version can continue to work with the updated version, that is a *backward-compatible* change. When software that works with the updated version of the component also works with the original version, the changes are said to be *forward-compatible*.

For example, an application uses an existing library, which is updated to a newer version. If the application continues to behave in the same manner with new library, then the new library has maintained a backward-compatible relationship, that is, **new code works with old code.**

A forward-compatible relationship is a little more tricky to achieve. If an earlier version of a library replaces an existing library and the application continues to behave in the same manner, the earlier version of the library is said to have a forward-compatible relationship, that is, **old code works with new code.**

Backward compatibility is typically the primary goal when making incremental releases of a component, with forward compatibility a desirable extra. Some changes cannot be forward-compatible, such as bug fixes, which by their nature do not work "correctly" in releases prior to the fix.

Source and binary compatibility

If a change is made to a component and its dependent components can recompile against it without the need to make any changes, it can be said to be a *source-compatible* change. An example of a source-compatible change is a bug fix to the internals of an exported function, which does not require a change to the function declaration itself (that is the component's interface).

On Symbian OS, a typical *source-incompatible* change involves modifying the internals of a member function to give it the potential to leave when previously it could not do so. To adhere strictly to the naming convention, the name of the function must also be modified by the addition of a suffixed `L` (see Chapter 5).

A source-compatible change does not mean the dependent components do not need to be recompiled, just that they do not need to be modified to be compiled successfully.

Binary compatibility is achieved when one component, dependent on another, can continue to run without recompilation or re-linking after the component on which it depends is modified. The compatibility extends across compilation and link boundaries.

One example of a binary-compatible change is the addition to a class of a public, non-virtual function which is exported from the library with an ordinal (see Chapter 10) and which comes after the previous set of exported functions. A client component which was dependent on the original version of the library is not affected by the addition of a function to the end of the export list of the library, so the change is binary-compatible and backward-compatible.

If the addition of a new function to the class causes the ordinals of the exported functions to be reordered, the change is not binary-compatible, although it continues to be source-compatible. Dependent code must be recompiled, otherwise it would use the original, now invalid, ordinal numbers to identify the exports. Section 10.1 discusses in more detail which functions should be exported from a DLL.

Class-level and library-level compatibility

Maintaining compatibility at a *class level* means ensuring that, among other things, methods continue to have the same semantics as were initially documented, no publicly accessible data is moved or made less accessible, and the size of an object of the class does not change (at least, not when a client is responsible for allocating memory for it). To maintain *library-level* compatibility, ensure that the API functions exported by a DLL are at the same ordinal and that the parameters and return values of each are still compatible.

Exam Essentials

- Demonstrate an understanding of source, binary, library, semantic and forward/backward compatibility

16.2 Preventing Compatibility Breaks – What Cannot Be Changed?

Critical Information

The size of a class object must not change

Changing the size of a class object, for example by adding or removing data, will cause a binary-compatibility break, unless it can be guaranteed that:

- The class is not externally derivable, that is, a constructor is not exported from the DLL which defines it
- The only code that allocates an object resides within the component/DLL being changed, or it has a non-public constructor that prevents it from being created on the stack
- The class has a virtual destructor (see Section 3.3).

The size of memory required for an object to be allocated on the stack is determined for each component at build time. To change the size of an object would affect previously compiled client code, unless the client is guaranteed to instantiate the object only on the heap, say by using a `NewL()` factory function.

Additionally, access to data members within an object occurs through an offset from the `this` pointer. If the class size is changed, say by adding a data member, the offsets of the data members of derived classes are rendered invalid.

To ensure that an object of a class cannot be derived or instantiated except by members (or friends) of the class, it should have private, non-inline and non-exported constructors. It is not sufficient simply not to declare any constructors for the class, since the compiler will then generate an implicit, public default constructor (see Chapter 2). If a class needs a default constructor, then it should be defined as private and implemented in the source, or at least, not inline where it is publicly accessible. All Symbian OS C classes derive from `CBase`, which defines a protected default constructor and prevents the compiler from generating an implicit version (see Chapter 4).

If something is accessible, it must not be removed

If something is removed from an API which is used by an external component, that component's code will no longer compile against the API (a break in source compatibility) nor run against its implementation (a break in binary compatibility).

At an API level, do not remove any:

- Externally visible classes
- Functions
- Enumerations
- Values within an enumeration
- Global data (such as string literals or constants).

At a class level, do not remove any:

- Methods
- Member data.

Private and protected member data should not be removed, because this will change the size of the resulting object.

Accessible member data must not be rearranged

Simply rearranging the order of member data can cause problems to client code that accesses that data directly, as the offset of the member data from the object's `this` pointer will be changed.

Do not change the position of member data in a class if that data is:

- Public (or protected, if the client can derive from the class)
- Exposed through public or protected inline methods (which will have been compiled into client code).

This rule also means that the order of the base classes from which a class multiply inherits cannot be changed without breaking compatibility, because this order affects the overall data layout of the derived object.

Exported functions must not be reordered

Each exported API function is associated with an ordinal number, which is used by the linker to identify the function. The function ordinals are stored in the module definition (`.def`) file (see Chapter 10).

If the `.def` file list is reordered, say by adding a new export within the list, the ordinal number values will change and previously compiled code will be unable to locate the correct function. For example, adding a new function at the start of the list will shunt all the ordinals up one; thus any component using ordinal `4` would be now be looking at what was previously ordinal `3`.

This change breaks binary compatibility. To avoid this, a new export should always be added to the end of the `.def` file, which assigns it a new, previously unused, ordinal value.

Virtual functions of externally derivable classes must not be added, removed or modified

If a class is externally derivable (that is, if it has an exported or inlined, public or protected constructor), adding, removing or modifying virtual functions will break compatibility.

If a derived class defines its own virtual functions, these will be placed in the virtual function table directly after those defined by the base class (see Chapter 2). If a virtual function is added or removed in the base class, there will be a change in the `vtable` position of any virtual functions defined by a derived class (see Chapter 3). Thus any code that was compiled against the original version of the derived class will now be using an incorrect `vtable` layout, breaking binary compatibility.

The following modifications of virtual functions will also break compatibility:

- Changing the parameters
- Modifying the return type
- Changing the use of `const`.

However, changes to the internal operation of the function, for example bug fixes, do not affect backward compatibility.

Virtual functions must not be reordered

Although this is not stated in the C++ standard, the order in which virtual member functions are specified in the class definition can be assumed be the only factor which affects the order in which they appear in the virtual function table. Therefore, this order should not be changed, since client code compiled against an earlier version of the virtual function table will call what has become a completely different virtual function.

Virtual functions that were previously inherited should not be overridden

Overriding a virtual function that was previously inherited alters the virtual function table of the base class. As existing client code is compiled against the original `vtable` it will continue to access the inherited base-class function rather than the new, derived version.

This leads to inconsistency between callers compiled against the original version of the library and those compiled against the new version. Although it does not strictly result in incompatibility, this is best avoided.

For example, a client of `CSiamese` version 1.0 calling `SleepL()` invokes `CCat::SleepL()`, while clients of version 2.0 invoke `CSiamese::SleepL()`:

```
class CCat : public CBase // Abstract base class
  {
public:
  IMPORT_C virtual ~CCat() = 0;
public:
  IMPORT_C virtual void PlayL(); // Default implementation
  IMPORT_C virtual void SleepL(); // Default implementation
protected:
  CCat();
  };
```

```
class CSiamese : public CCat // Version 1.0
  {
public:
  IMPORT_C virtual ~CSiamese();
public:
  // Overrides PlayL() but defaults to CCat::SleepL()
  IMPORT_C virtual void PlayL();
  // ...
  };

class CSiamese : public CCat // Version 2.0
  {
public:
  IMPORT_C virtual ~CSiamese();
public:
  // Now overrides PlayL() and SleepL()
  IMPORT_C virtual void PlayL();
  IMPORT_C virtual void SleepL();
  // ...
  };
```

The documented semantics of an API should not be modified

Changing the documented behavior of a class or global function, or the meaning of a constant, may break compatibility, regardless of whether source and binary compatibility are maintained.

As a very simple example, consider a class which, when supplied with a data set, returns the average value of that data. If the first release of the `Average()` function returned the arithmetic mean value, the second release of `Average()` should continue to do so and not return a median value, or some other interpretation of an average.

Default arguments specified in header files are compiled into client code. Although a change made to a default argument doesn't break binary or source compatibility, the client must be recompiled to pick up the change. A client using the old default argument could get an unexpected return value, which would be a problem because the behavior of a function also forms part of its interface.

Use of `const` should not be removed

The semantics of "`const`" should not be removed, since this will be a source-incompatible change. This means that the `const`ness of a parameter, return type or method should not be removed.

Parameters passed by value must not be changed to pass them by reference, or vice versa

If function input parameters or return values were originally passed by value, changing them to pass by reference breaks binary compatibility. This is also true for changing references to values.

When a parameter is passed by value, the compiler generates a stack copy and passes it to the function (See Chapter 2). However, if the function signature is changed to accept the parameter by reference, a word-sized reference to the original object is passed to the function instead.

The stack frame usage for a pass-by-reference function call is significantly different from that for a pass-by-value function call, causing binary incompatibility.

```
class TColor
  {
  ...
private:
  TInt iRed;
  TInt iGreen;
  TInt iBlue;
  };

// version 1.0
// Pass in TColor by value (12 bytes)
  IMPORT_C void Fill(TColor aBackground);

// version 2.0 -  binary compatibility is broken
// Pass in TColor by reference (4 bytes)
  IMPORT_C void Fill(TColor& aBackground);
```

Exam Essentials

- Recognize which attributes of a class are necessary for a change in the size of the class data not to break compatibility
- Understand which class-level changes will break source compatibility
- Understand which class-level changes will break binary compatibility
- Understand which library-level changes will break binary compatibility
- Understand which function-level changes will break binary and source compatibility
- Differentiate between derivable and non-derivable C++ classes in terms of what cannot be changed without breaking binary compatibility

16.3 What Can Be Changed Without Breaking Compatibility?

Critical Information

An API may be extended

Classes, constants, global data or functions may be added without breaking compatibility. Likewise, a class can be extended by the addition of static member functions or non-virtual member functions (but not virtual member functions, as described in Section 16.2).

The ordinals for exported functions must be added to the bottom of the module definition file (`.def`) export list to avoid re-ordering the existing functions.

The private internals of a class may be modified

Changes to private and protected methods that are neither exported nor virtual do not break client compatibility.

However, the functions must not be called by externally-accessible inline methods, since the call inside the inline method would be compiled into external calling code and would be broken by an incompatible change to the internals of the class.

Changes to private member data are also permissible, unless they result in a change to the size of the object or move the position of public or protected data in the object (exposed directly, through inheritance or through public inline accessor methods).

Access specification may be relaxed

The C++ access specifier (`public`, `protected`, `private`) doesn't affect the layout of a class and can be relaxed without affecting the data order of the object. The position of member data in an object is determined solely by the order of specification in the class definition.

Changing the access specification to a more restricted form, for example from public to private, means that the member data becomes invisible to external clients when previously it was visible. This breaks source compatibility, but not binary compatibility.

Pointers may be replaced with references and vice versa

Changing from a pointer to a reference parameter or return type (or vice versa) in a class method does not break binary compatibility (but does break source compatibility). This is because references and pointers can be considered to be represented in the same way by the C++ compiler that is one machine word.

The names of exported non-virtual functions may be changed

Symbian OS is linked purely by ordinal and not by name and signature. This means that it is possible to make changes to the name of exported functions and retain binary, though not source, compatibility.

The input may be widened and output narrowed

Input can be made more generic, or *widened*, as long as input that is currently valid retains the same interpretation. For example, a function can be modified to accept a less derived pointer, and extra values can be added to an enumeration (as long as it is extended rather than re-ordered, which would change the original values).

Output can be made less generic ("*narrowed*") as long as any current output values are preserved. For example, the return pointer of a function can be made more derived as long as the new return type applies to the original return value.

For multiple inheritance, say, a pointer to a class is unchanged when it is converted to a pointer to the first base class in the inheritance declaration order. That is, the layout of the object follows the inheritance order specified.

The `const` specifier may be applied

It is acceptable to change non-`const` parameters, return types or the `this` pointer to be `const` in a non-virtual function, as long as the parameter is no more complicated than a reference or pointer. This is because it is possible to pass non-`const` parameters to `const` functions or those that take `const` parameters. In effect, this is an extension of the "input widening" guideline described above.

Exam Essentials

- Understand which class-level changes will not break source compatibility
- Understand which class-level changes will not break binary compatibility
- Understand which library-level changes will not break binary compatibility
- Understand which function-level changes will not break binary and source compatibility
- Differentiate between derivable and non-derivable C++ classes in terms of what can be changed without breaking binary compatibility

16.4 Best Practice – Designing to Ensure Future Compatibility

Critical Information

Functions should not be inline

An inline function is compiled into the client's code, which means that a client must recompile its code in order to pick up a change to an inline function. When using private inline methods within a class, they must not be accessible externally, and should be implemented in a file that is accessible only to the code module in question.

Using an inline function increases the coupling between a component and its dependents, which should generally be avoided.

No public or protected member data should be exposed

The position of data is fixed for the lifetime of the object if it is externally accessible, either directly or through derivation. More flexibility is achieved by encapsulating member data privately and providing non-inline accessor functions where necessary.

Derived virtual functions should be stubbed

This is a defensive programming technique where a derived class overrides all the base-class virtual functions regardless of whether they are needed in the first release (where there is no need to modify the functions beyond what the base class supplies, the overridden implementation should simply call the base-class function). This allows the functions to be extended in future releases.

The earlier example of `CSiamese`, deriving from `CCat`, inherited the default implementation of the `CCat::SleepL()` virtual method in version 1.0 but overrode it in version 2.0. The sample code below shows how to avoid this by overriding both virtual functions in version 1.0, although `CSiamese::SleepL()` simply calls through to `CCat::SleepL()`.

```
Class CCat : public CBase // Abstract base class
  {
public:
  IMPORT_C virtual ~CCat() = 0;
public:
  IMPORT_C virtual void EatL(); // Default implementation
  IMPORT_C virtual void SleepL();// Default implementation
  // ...
  };

class CSiamese : public CCat // Version 1.0
  {
  IMPORT_C virtual ~CSiamese();
public:
  IMPORT_C virtual void EatL(); // Overrides base class functions
  IMPORT_C virtual void SleepL();
  // ...
  };

// Function definitions not relevant to the discussion have been
// omitted
void CSiamese::EatL()
  {// Overrides base class implementation
  ... // Omitted for clarity
  }

void CSiamese::SleepL()
  {// Calls base class implementation
  CCat::SleepL();
  }
```

"Spare" member data and virtual functions should be provided from the outset

Another practical defensive programming technique is to add at least one reserve exported virtual function where there is the possibility for future expansion. This provides the means to extend the class without disrupting the `vtable` layout of classes.

Also, reserving at least four extra bytes of private member data in classes is a good future-proofing technique. This reserved data can be used as a pointer to extra data as it is required. However, if the class is unlikely to require later modification, extra memory should not be reserved to avoid wasting limited memory resources.

Exam Essentials

- Recognize best practice for maintaining source and binary compatibility
- Recognize the coupling arising from the use of inline functions and differentiate between cases where it will make maintaining binary compatibility more difficult and where it will be less significant

References

[Stichbury 2004 Chapter 18]

Bibliography

Alexandrescu, Andrei (2001) *Modern C++ Design: Generic programming and design patterns applied*, Addison-Wesley Professional.

Ambler, S. W. (2001) *The Object Primer: Agile model-driven development with UML*, 2nd Edition, Cambridge University Press.

Babin, S. (2005) *Developing Software for Symbian OS: An introduction to creating smartphone applications in C++*, John Wiley & Sons.

Dewhurst, S. C. (2005) *C++ Common Knowledge: Essential intermediate programming*, Addison-Wesley Professional.

Gamma, E., Helm, R., Johnson, R. and Vlissides, J. (1995) *Design Patterns: Elements of reusable object-oriented software*, Addison-Wesley Professional.

Harrison, R. (2004) *Symbian OS C++ for Mobile Phones Volume 2*, John Wiley & Sons.

Heath, C. (2006) *Symbian OS Platform Security: Software development using the Symbian OS security architecture*, John Wiley & Sons.

Jipping, M. J. (2002) *Symbian OS Communications Programming*, John Wiley & Sons.

Meyers, S. (2005) *Effective C++: 55 specific ways to improve your programs and designs*, 3rd Edition, Addison Wesley.

Morris, B. (2006) *The Symbian OS Architecture Sourcebook: Design and evolution of a mobile phone OS*, John Wiley & Sons.

Sales, J. (2005) *Symbian OS Internals: Real-time kernel programming*, John Wiley & Sons.

Stallings, W. (2004) *Wireless Communications & Networks* (2nd Edition), Prentice Hall.

Stichbury, J. (2004) *Symbian OS Explained: Effective C++ programming for smartphones*, John Wiley & Sons.

Stroustrup, B. (2000) *The C++ Programming Language* (Special 3rd Edition), Addison-Wesley Professional.

Sutter, H. (1999) *Exceptional C++,* Addison-Wesley Professional.

Sutter, H. and Alexandrescu, A. (2004) *C++ Coding Standards: 101 rules, guidelines and best practices,* Addison-Wesley Professional.

Online References

Symbian Developer Library ***www.symbian.com/developer***

Other online references are noted in the chapters to which they are relevant. However, since URLs can change over time, we will also keep an up-to-date list of useful online references on the Meme Education website (***www.meme-education.com***). Please contact us from there if you find one of the online resources listed in the book unavailable and we will do our best to find a replacement.

Index

Learn more with Symbian Press

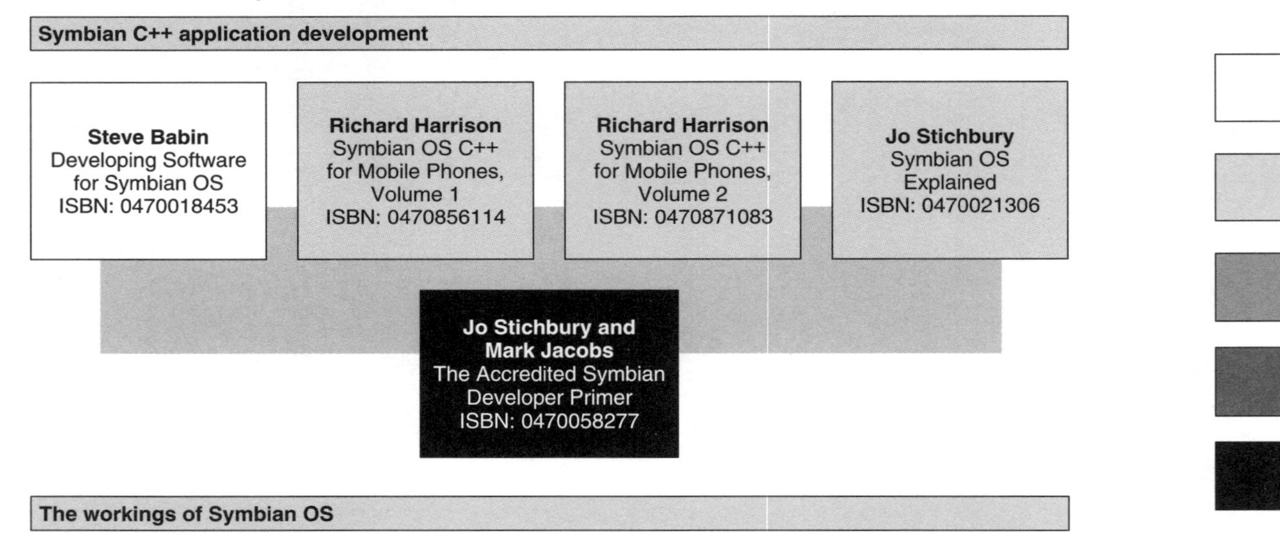

Symbian C++ application development
Steve Babin
Developing Software for Symbian OS
ISBN: 0470018453
Richard Harrison
Symbian OS C++ for Mobile Phones, Volume 1
ISBN: 0470856114
Richard Harrison
Symbian OS C++ for Mobile Phones, Volume 2
ISBN: 0470871083
Jo Stichbury
Symbian OS Explained
ISBN: 0470021306
Jo Stichbury and Mark Jacobs
The Accredited Symbian Developer Primer
ISBN: 0470058277
The workings of Symbian OS
Jane Sales
Symbian OS Internals
ISBN: 0470025247
NEW
Ben Morris
The Symbian OS Architecture Sourcebook
ISBN: 0470018461
Introductory
For developers wishing to extend their knowledge
Advanced topics
Reference
You are here